About Island Press

Since 1984, the nonprofit organization Island Press has been stimulating, shaping, and communicating ideas that are essential for solving environmental problems worldwide. With more than 1,000 titles in print and some 30 new releases each year, we are the nation's leading publisher on environmental issues. We identify innovative thinkers and emerging trends in the environmental field. We work with world-renowned experts and authors to develop cross-disciplinary solutions to environmental challenges.

Island Press designs and executes educational campaigns, in conjunction with our authors, to communicate their critical messages in print, in person, and online using the latest technologies, innovative programs, and the media. Our goal is to reach targeted audiences—scientists, policy makers, environmental advocates, urban planners, the media, and concerned citizens—with information that can be used to create the framework for long-term ecological health and human well-being.

Island Press gratefully acknowledges major support from The Bobolink Foundation, Caldera Foundation, The Curtis and Edith Munson Foundation, The Forrest C. and Frances H. Lattner Foundation, The JPB Foundation, The Kresge Foundation, The Summit Charitable Foundation, Inc., and many other generous organizations and individuals.

The opinions expressed in this book are those of the author(s) and do not necessarily reflect the views of our supporters.

Tech to Table

Tech to Table

25 INNOVATORS REIMAGINING FOOD

Richard Munson

ISLANDPRESS | Washington | Covelo

Library of Congress Control Number: 2021932940

All Island Press books are printed on environmentally responsible materials.

Manufactured in the United States of America
10 9 8 7 6 5 4 3 2 1

Keywords: ag-tech start-up, aquaculture, blockchain, cell-cultured meat, CRISPR, disruptive technologies, farm drones, food waste, hydroponics, insect farming, lab-grown meat, mechanical harvesting, methane blocker, pesticide alternatives, plant-based protein, precision agriculture, soil probiotics, 3D printing meals, venture capitalists, vertical farming

To Daniel, Dana, and Ryan,
the next generation;

and in memory of Lauren

"We know more about the movement of celestial bodies than we know about the soil underfoot."

—Leonardo da Vinci

Contents

PART 3. CURTAIL POISONS   121

PART 4. NOURISH PLANTS   143

PART 5. CUT CARBON   183

The Rise of Innovators

The rolling farm fields of Iowa and the sleek glass-and-steel office parks of California's Silicon Valley could hardly present two more different pictures of the American economy. Yet today, these worlds are colliding in the realm of food and agriculture, disrupting an industry well known for its conventionalism. Within just the past decade, new technologies have begun to redefine what it means to farm, and even the nature of food. Agricultural-technology, or ag-tech, start-ups, twenty-five of which are profiled in the chapters that follow, have soared by 80 percent annually since 2012, attracting more than $30 billion in direct investment in 2020.[1]

The scope of these innovators startles. Serial entrepreneurs grow crops year-round and without soil on stacks within large urban warehouses, moving farming closer to consumers and from the horizontal to the vertical. Medical doctors make meats from plants and cultured stem cells, delivering proteins without slaughtering animals. Engineers deploy drones, ground-based sensors, and precision controls to track and apply the water and nutrients needed by individual plants throughout massive fields. Roboticists send autonomous machines to pick fruits and pluck weeds, reducing drudgery and curtailing the need for poisonous herbicides. Biologists, confronting and tackling climate change, edit genes so plants withstand droughts or sequester carbon. Chemists devise probiotics that nourish microbes and roots, avoiding the need for synthetic fertilizers. Bankers compile sophisticated ledgers that trace crops from seeds to tables, empowering consumers with information about their food and allowing farmers to benefit from their products' special attributes.

These diverse innovators are supplanting the long-standing debate between Big Ag[2] and return-to-husbandry advocates. While industrial agriculture focuses on chemicals and monocultures, reformers call for organic and regenerative cultivation, winter cover crops, less plowing, and more composting, yet neither camp has proven capable of providing sufficient supplies of nutritious food without environmental damage. Malnutrition and pollution flourish under Big Ag, and despite decades of effort, organic farming accounts for less than 1 percent of US croplands, and broad adoption of regenerative practices has failed to materialize.[3] Entrepreneurial innovators represent a third path, an alternative that advances disruptive technologies not just to feed a growing population but also with the express goal of sustainability.

The individuals driving these developments are often as unconventional as their inventions. They tend to be outsiders, people who have worked in technology rather than agriculture. Most are young, idealistic, but also aggressive competitors thrilled to challenge the status quo. They look at food and farms from fresh perspectives and see opportunities.

The twenty-five innovators profiled in these pages come from a variety of professional, ethnic, and geographic backgrounds. Their specific motivations and endgames differ. Yet each is part of a widespread revolution in how we farm and what we eat. Some of their inventions have received full-blown media attention, so you may have already tasted plant-based burgers that are indistinguishable from beef, but others have yet to reach widespread markets. Have you sipped vodka made from CO_2 or dined on food made by a 3D printer?

You might ask yourself, Do I want to? In other words, should we trust these new technologies? My answer is, not blindly. The efforts of the green revolution technologists of the 1950s and 1960s increased crop yields and reduced worldwide hunger, but they poisoned our lands and water, decreased food diversity, and limited competition. Today's innovations must be judged by their long-term sustainability (their ability to protect the planet) and equity (their ability to provide nutritious food to a growing population). They must show declines in the number of malnourished individuals as well as agriculture's greenhouse-gas emissions. Through measurements of soil, water, and air quality, they must demonstrate transparently that they meet "the needs of the present without compromising the ability of future generations to meet their own needs."[4]

There is of course no guarantee that the innovations covered here, or the new ones that will be hitting the market in the coming years, will meet those criteria. But unlike the agricultural technologists of the 1950s and 1960s, today's entrepreneurs do have the benefit of hindsight, of seeing how technologies that disrupt natural systems can backfire. A common theme throughout these profiles is mimicking nature rather than running roughshod over it. These innovators are more likely to be guided by inconvenient truths than better living through chemistry; they have put sustainability front and center in ways that would not have occurred to earlier generations of scientists and technologists.

Yet the environmental and health consequences of our industrial food system have been known for decades. So why now? Why are so many agricultural innovators suddenly attracting investments and capturing markets? The short answer is opportunity. First, the confluence of technological advances—including computers, sensors, robots, and machine learning—allow fast-moving disruptors to thrive, particularly in an agricultural sector that has been slow to adopt innovation. Second, visionaries and their financiers increasingly believe they can outcompete Big Ag's slow-moving oligopolies. Finally, and most importantly, the sheer size of our challenges—to double food availability and slash pollution—demands creative thinkers and actors.

Before delving into the personalities and products of those actors, it's worth examining the factors that brought us to this moment.

Confluence of Technologies

Agriculture remains the least digitalized of all businesses,[5] reflecting a cautiousness that makes the farm and food sectors slow to respond to population and environmental challenges—but also vulnerable to creative disruption. Even though GPS mapping has been available for decades, only about half of the large corn and soybean farmers in the United States deploy such systems in their tractors, less than 20 percent of them utilize variable rate technology to target their fertilizer and herbicide spraying, and very few yet take advantage of CRISPR gene-editing technology, robotics, or other modern innovations.

Yet we live in an era of rapid technological advances across numerous economic sectors. Sophisticated sensors collect enormous quantities of high-resolution data, which high-performance computers decipher to deliver real-time insights and predictions. Autonomous machines perform complex tasks with speed and precision, while gene editors enable organisms to retard chronic diseases. According to Eric Schmidt, former CEO of Google and Novell, this radical convergence of data, leading-edge computation, and advanced engineering creates a "super evolution" that will "fundamentally, irrevocably transform" wide-ranging industries.[6] He claims that such innovations allow start-ups "to advance faster than incumbents," resulting in "extremely agile, powerful companies."[7]

Although farming long has been a technology straggler, this "super evolution" is reaching the agricultural sector, spurred by competitive entrepreneurs seeing opportunities to profit through innovation.

Agricultural technology has become "sexy" and is no longer a fringe investment category, says Seana Day, a partner with Better Food Ventures. "Lots of new money is flowing into food and farm innovation." The investor adds that "we're still in the first innings associated with agriculture's basic digitalization, but opportunities abound."[8]

Outcompeting Oligopolies

Food and farm entrepreneurs believe agriculture's problems intensified with the rise of stodgy corporate oligopolies—often known as Big Ag. From 1988 to 2015, four biotech companies increased their combined share of the corn seed market from 50.5 percent to 85 percent.[9] The four largest meat-packers raised their stakes in cattle slaughtering to 85 percent,[10] just three megabuyers dominate 87 percent of the corn market,[11] and the four biggest pesticide manufacturers control 57 percent of their industry. This trend also goes for grocery retailers and the makers of livestock pharmaceuticals and farm machinery, which have increased their consolidation significantly since the 1990s.[12]

These corporate concentrations—resulting from mergers and vertical integrations—squeeze farmers with higher charges for seeds, machinery, and fertilizers and then squeeze them again when growers try to sell their crops. Big Ag accelerates a focus on monocultures and reduces the diversity of food options. Vested in the status quo, these giant corporations mostly embrace and defend current farm products and practices.

Big Ag invests little in research and development, offering an opening to science- and technology-focused innovators. While the health-care industry spends 21 percent of its budget on research and development—and computing devotes 25 percent, and automotive commits

16 percent—the global food companies allocate less than 1 percent. Put simply, industrial agriculture focuses far more on protecting current practices than on creating novel ones.

Several Big Ag corporations do develop new offerings. Koch Agronomic Services, for instance, introduced nitrification inhibitors that ensure more nitrates nurture plant roots. Yet such an advancement merely tweaks existing approaches and enables agricultural giants to sell more of their conventional products.

A few oligopolists buy up creative start-ups, whether to contain competitors, to advance lines of invention, or to spur new products. Deere & Company, for instance, bought Blue River Technology (profiled in chapter 14) and its precision weeding devices for $305 million. DuPont purchased Granular, a producer of farm-management software, for $300 million, and Monsanto spent $1 billion to control the Climate Corporation, whose app collects farm-field data. Yet this book's profiles suggest that independent entrepreneurs still control and drive most of today's agricultural advances.

Big Ag corporations tend to oppose disruptive innovation and hire well-connected lobbyists, lawyers, and advertisers to challenge alternatives that threaten their status quo. To date, those efforts have had little impact. Smart money is going to ag-tech innovators, who in 2019 launched eight hundred start-up firms and attracted more than $17 billion in direct venture investment, up 43 percent from the previous year and representing a fivefold increase since 2012.[13] The Switzerland-based bank UBS predicts ag-tech's sales will climb to $700 billion by 2030.[14]

Large corporations, by their very nature, usually fail to develop disruptive technologies, even if their managers are early to spot such openings. According to Clayton Christensen, a former Harvard Business School professor, to expect executives at big companies "to do something like nurturing disruptive technologies—to focus resources on proposals that customers reject, that offer lower profit, that underperform existing technologies and can only be sold in insignificant markets—is akin to flapping one's arms with wings strapped to them in an attempt to fly."[15]

Such flapping opens opportunities for the nimble and aggressive, those innovators trying creative ways to feed a growing population and protect a fragile planet.

Challenges

Perhaps Big Ag is resistant to innovation because, in some ways, the current food system functions quite well. Walking through any supermarket, at any time of the year, we expect to choose from a cornucopia of in- and out-of-season fruits and vegetables, as well as an assortment of cereals, meats, and dairy products. In fact, crop yields have risen fivefold since 1940. At the same time, food's share of an average American's budget has fallen by half since the 1950s, to 10 percent, meaning more people can afford meals.

Yet agriculture is in trouble, causes trouble, and needs disruption. Industrialized farming, according to some calculations, is the single biggest cause of climate change:[16] the livestock sector alone emits more, probably far more, greenhouse gases than all cars, trains, boats, and airplanes combined.[17] Plowing exposes farmlands to rain and wind erosion, leading to the loss of thirty soccer fields' worth of soil every minute,[18] costing $44 billion each year.[19] Irrigating crops devours almost 80 percent of the planet's fresh water, depleting aquifers and acidifying soils.

In addition to the harm done to the environment by raising farm animals, sowing seeds, and watering plants is the harm done by enhancing these processes with chemicals. Synthetic additives exterminate beneficial microbes and shrink the land's fertility by almost one-half of a percentage point each year, an unsustainable trend with profound social, economic, and political implications. Toxic pesticides attack biodiversity, reducing the natural sustainability of all life-forms. In the United States, the spraying of pesticides annually kills seventy million birds and billions of bees and other useful bugs; just over the past three decades, the planet lost 75 percent of its flying insects.[20] Meanwhile, the runoff and leakage of chemical fertilizers and pesticides cause 70 percent of US water pollution, poison private wells, and form dead zones in the oceans and lakes where no aquatic life can survive.[21]

These environmental problems are poised to become worse. A study led by the World Resources Institute found that "if today's levels of production efficiency were to remain constant through 2050, then feeding the planet would entail clearing most of the world's remaining forests, wiping out thousands more species, and releasing enough [greenhouse-gas] emissions to exceed the 1.5°C and 2°C warming targets enshrined in the Paris Agreement—even if emissions from all other human activities were entirely eliminated."[22]

This damage is often cast as the cost of feeding almost ten billion people by 2050—two billion more than today, or the equivalent of adding the population of one and one-half Chinas. In truth, the world's farmers already produce enough calories to nourish ten billion humans, allowing approximately 2,700 calories per day per person. It's just that not all those calories go to humans; instead, almost 40 percent is lost as wasted food, roughly 30 percent feeds cattle and other animals, and 5 percent is used to brew biofuels and industrial products.[23] Critics call out the "feeding the world" justification as "a club crafted and wielded by agribusiness to scare us into supporting their agenda" of chemicals, irrigation, and monocultures.[24]

Vocal reformers assert that a return to preindustrial husbandry will protect the environment and provide healthier foods, yet organic and regenerative agriculture remains a tiny portion of the food and farm sectors. Usually facing higher management costs and lower yields, these cultivators confront the challenge of scaling up production.

Despite Big Ag's focus on growing continually more, we also confront a fundamental food problem not of quantity but of quality, equity, and distribution. Today, 815 million people go hungry and another 2 billion are overweight or obese.[25] Fully 20 percent of worldwide deaths—as well as debilitating diseases such as diabetes, cancer, and osteoporosis—result from bad nutrition; according to one report, "unhealthy diets now pose a greater risk to morbidity and mortality than unsafe sex, alcohol, drug, and tobacco use combined."[26] Malnutrition seems ubiquitous and not limited to developing nations; the United States' obesity rate approaches 40 percent,[27] and approximately 14 percent of American households lack sufficient food.[28]

Some nutritionists argue that we can solve malnutrition "simply" by changing consumers' food preferences, yet decades of admonitions to eat less meat and sugar have achieved little impact. While poverty remains a key cause of nutritional deficiencies, wealthy and poor alike often eat too much unhealthy food. Such habits—which retard cognitive development and academic performance—particularly cripple younger generations.

Growing prosperity has in fact generally worsened the environmental and health problems created by industrial farming, particularly in India and China, where consumers desire more protein-rich meats, eggs, and dairy products. That trend could vastly increase carbon pollution if we slash forests to clear farmlands on which to plant additional corn to feed more cattle, pigs, and chickens.

One could reasonably think that even though it is bad for the environment and human health, industrial agriculture is at least profitable. That is true for giant agribusinesses, not so much for the average farmer. "The cost of farming is soaring," complains Hank Scott, a third-generation cultivator who grows mostly sweet corn and fresh produce outside Mount Dora, Florida. "It's a tough business."[29] It needs innovation.

Scott calls farming a "delicate balancing act" of trying "to grow more

produce on less land on less margins, and ever-increasing cost of production, and still make a profit." He also fears the unexpected. It is hard,
Scott says, "to stay positive and upbeat when you see six weeks of hard
work and investment destroyed by a one-hour weather event."[30]

Small and midsize producers confront most of the financial stress.
Over approximately the past two decades, these farms—with gross
annual receipts under $350,000—saw their share of US farm pro-
duction fall from 46 percent to under 25 percent.[31] And more than
40 percent of midsize farms—with gross incomes between $350,000
and $1 million—face profit margins below 10 percent, making them
highly vulnerable to economic setbacks.[32] Farm bankruptcies rose in
2019 by 24 percent.[33] Stresses have been particularly stark for Black
farmers; reflecting "decades of racial violence and unfair lending and
land ownership policies," the share of farms run by African Americans
fell from 14 percent in 1920 to less than 2 percent in 2020.[34]

Scott sadly (and angrily) admits that many people view farmers as
enemies. "Ever since pesticides, herbicides, fungicides and fertilizers
were considered evil," he complains, farmers have been "portrayed as
being too dumb to know how to be good stewards of their land and
water."[35] He blames the media for not recognizing the benefits of farming. He argues that growers deserve far more respect for their hard work
and life-giving crops, and he cites a quote by chef Alice Waters: "We all
need to support the people who are taking care of the land. They are to
be treasured. We all need to get to know them."[36]

Still, Scott acknowledges that many of the needed changes will come
from outside conventional agriculture, from entrepreneurs comfortable
with the fast pace, financing, and fresh thinking associated with Silicon
Valley and other tech centers. That said, Scott appropriately warns that
innovators must engage farmers, asking them what they need rather
than telling them what to do. Although nervous about both current
financial strains and future technology disruptions, Scott hopes new

players will fuel rural revitalization, bring more options to today's farmers, and offer additional on-the-land opportunities for their daughters and sons.[37]

Disruptors

Agriculture is the world's largest industry—valued at $15 trillion, approximately 17 percent of the world's economy—and what we eat defines our culture and identity; food can be either medicine or poison.[38] Transforming farms and food, therefore, is no small task, and many in the Big Ag and organic camps dismiss the potential for disruptive entrepreneurs. Representatives of Oliver Wyman, a global management-consulting firm, wrote, "The reality is that very little innovation has taken place in the [agricultural] industry of late."[39]

I disagree, and this book highlights entrepreneurs bringing innovation to agriculture. Food has become the future's crucial battleground. The EAT-*Lancet* Commission, composed of thirty-seven researchers from sixteen countries, summarized the stakes: "Food is the single strongest lever to optimize human health and environmental sustainability on Earth."[40] One of the authors of the commission's report, Johan Rockström of Sweden's Stockholm Resilience Centre, reflected, "Humanity now poses a threat to the stability of the planet. [This requires] nothing less than a new global agricultural revolution."[41] Also pushing farm and food changes are demographic shifts, particularly the rise of millennials—the huge cohort born in the 1980s and 1990s—who demand healthy and transparent ingredients. According to one researcher, their purchasing power caused 90 percent of Big Ag's top consumer packaged goods to lose market share and clean-label alternatives to grow sixfold.[42]

Transforming agriculture will require private-sector players who recognize the power of markets and who can attract investments, ramp up food production, and streamline distribution. Pleas from chefs and nutritionists are not sufficient; if they were, consumers would have stopped eating too few plants and too much sugar long ago. Similarly, it will take more than studies by academics and lobbying by environmentalist activists to curb agriculture's pollution and greenhouse-gas emissions, which continue to rise.

The profiles that follow—entrepreneurs transforming how we grow and ranch—are organized according to five overarching challenges facing farms and food. Of equal importance, these goals are to deliver more proteins, reduce food waste, curtail poisons, nourish plants, and cut carbon pollution.

Beyond this book, we will continue to have heated ethical arguments about food choices: whether, for instance, to eat meats, consume genetically modified foods, or adopt vegetarian, vegan, keto, Paleo, Mediterranean, Atkins, or other diets. Big Ag and soil-health reformers—be they advocating regenerative or organic approaches—will continue to debate, even as neither camp has been able to supply sufficient and healthy food sustainably. My goal is to reveal the substantial private-sector innovation that is disrupting agriculture, that is providing alternative perspectives—and progress—on long-simmering food and farm challenges.

To concentrate on the key players behind emerging trends and technologies, though I acknowledge the important research done at universities, I have chosen to feature business builders rather than academics. Likewise, recognizing the critical impact of public policy, I have avoided profiling politicians and lobbyists and focused on entrepreneurs. While two portrayals address alternatives to industrialized fishing, I concentrate on innovations affecting land-based crops and livestock.

The next five parts tell stories about the astonishing pace and breadth of change within agriculture, and they introduce businesswomen and businessmen attracting capital and building markets to address farm and food challenges. Although these innovators may not all prosper over the long term, they are advancing the disruptive approaches needed to reverse environmental degradation while delivering nutritious choices to feed us all.

PART I

DELIVER PROTEINS

As early as 1932, Winston Churchill wrote an essay about future trends, predicting, "We shall escape the absurdity of growing a whole chicken in order to eat the breast or the wing, by growing these parts separately in a suitable medium."[1] Thirty-eight years later, *National Geographic* magazine imagined "meatless dishes tasting like chicken, beef, or ham."[2] And in the early 2000s, the National Aeronautics and Space Administration and New York's Touro College extracted and grew goldfish cells to produce self-replenishing proteins for astronauts on lengthy expeditions.[3]

In the past decade, meat not derived from the slaughter of animals has gone from a flight of fancy to real products as consumers have demanded, and innovators have developed, alternative ways to obtain dietary proteins, which are crucial for building and maintaining healthy tissues and bones. Marcus Johannes "Mark" Post, a professor of vascular physiology at Maastricht University, could arguably be called the father of one portion of this emerging field, having obtained funding from the Dutch government and Google cofounder Sergey Brin to fashion the first laboratory-cultured meat. The Netherlands-based researcher in 2013 biopsied from a cow a few high-quality cells that renewed

themselves when fed a mixture of essential ingredients, including vitamins, amino acids, sugars, and oxygen. That initial burger cost almost
$325,000 and took three lab technicians three months to grow the
twenty thousand muscle fibers within the five-ounce patty;[4] two years
later, entrepreneurs dropped that cost to less than $12. At the burger's
rollout, Post asserted, "The few cells that we take from this cow can turn
into 10 tons of meat."[5] Despite technological and marketing hurdles,
investors are optimistic and more innovators are joining this fast-growing field each year, with about sixty companies worldwide.[6]

Less futuristic than lab-grown meat, but perhaps more palatable to
eaters wary of the idea of cultured cells, plant-based protein is also on
the rise, with sales totaling $5 billion in 2019 and climbing by 158
percent in 2020.[7] The financial firm UBS predicts that the US market
will grow to $85 billion by 2030, and *Entrepreneur* magazine suggests
that plant-based meat "has officially reached 'global phenomenon' status."[8] In 2019, Impossible Foods began selling its plant-based burgers at
Burger King, QDOBA, and dozens of other restaurants, while Beyond
Meat in March 2020 formed three-year partnerships with McDonald's
and Yum! Brands, which includes the KFC, Pizza Hut, and Taco Bell
chains. In just one year, according to a 2020 report, "plant-based meat
went from something very few Americans had heard of to something
that 40 percent of us have tried."[9]

Some of the buzz for alternative meat results from Big Meat's problems. COVID-19 highlighted the conventional supply chain's vul-
nerabilities as outbreaks among workers closed slaughterhouses, dairy
farmers dumped thousands of gallons of milk, and pork production
fell by 50 percent. Farmworkers, who already faced low wages and lax
safety requirements, suffered disproportionately high rates of coronavirus cases; according to a special report in the *Washington Post*, "small
farmers, new farmers and farmers of color, struggling in the shadow
of Big Ag, have been disproportionately affected by the pandemic and

are often not eligible for federal relief."[10] The virus revealed industrial agriculture's lack of resilience and the unsustainability of its growing corporate concentration; the closure of a single facility often disrupted the entire food chain and led to higher prices and reduced availabilities.

The pandemic, conversely, opened opportunities for innovators, as evidenced by the Impossible Burger's expanded distribution in 2020 to 1,700 Kroger grocery stores. Seth Bannon of Fifty Years, a San Francisco–based seed-capital fund, believes the virus "will cause way more capital to flow into the [alternative protein] space, as the world wakes up to the very real threat of zoonotic diseases."[11] Barclays analysts in 2019 forecast that this market would grow by 1,000 percent over the next ten years, reaching $140 billion.[12]

Disease risk is of course just the latest in a long line of reasons we need alternatives to industrialized meat. Since Upton Sinclair's *The Jungle* was published in 1905, journalists have well documented the poor working conditions and animal abuses in the meat industry. Slaughterhouses have an injury rate exceeding 27 percent annually, the highest of any business,[13] while wages have fallen from $24 per hour in inflation-adjusted dollars in 1982 to less than $14 per hour in 2020.[14] No doubt the animals fare even worse: factory farms confine breeding pigs in gestation crates, inseminate dairy cows to keep them giving milk—killing male offspring for veal—and raise turkeys with breasts so large they topple over. Slaughterhouses in the United States kill about three hundred birds every single second of every day, and they grind to death the male baby chicks because they cannot lay eggs. Philosopher Yuval Noah Harari observes, "Judged by the amount of suffering it causes, industrial farming of animals is arguably one of the worst crimes in history."[15]

Even leaving aside animal cruelty, meat production is not an efficient use of natural resources. For each animal fed 100 calories of grain, we obtain only 3 calories from beef, 40 calories from cow's milk,

12 calories from chicken, 22 calories from eggs, and 10 from pork.[16] According to EarthSave, a California-based nonprofit, "it takes 2,500 gallons of water, 12 pounds of grain, 35 pounds of topsoil and the energy equivalent of one gallon of gasoline to produce one pound of feedlot beef."[17]

Livestock account for a large share of greenhouse-gas emissions, with independent calculations ranging from 14.5 percent to as high as 51 percent.[18] Most such emissions result from the ruminating of beef and dairy cattle (as well as sheep and goats), whose bathtub-size stomachs churn together microbes that lead to the animals burping up methane, a gas eighty-four times more potent than carbon dioxide in causing climate change.[19] If cattle were a country, they would be the third-largest greenhouse-gas-emitting nation, and their numbers are expected to double to 40 billion animals by 2050.

Meanwhile, feeding and medicating livestock presents its own crisis. Growing food for cows, pigs, sheep, and chickens requires almost one-third of the planet's arable land. And the overuse of antibiotics in factory farming, both to treat animals for disease and to fatten them up, is leading to the rise of superbugs. Meat and poultry production accounts for about 80 percent of US antibiotic use, while roughly 1.5 million people die each year from drug-resistant illnesses.[20]

The United Nations has called the raising of livestock "one of the top two or three most significant contributors to the most serious environmental problems, at every scale from local to global. . . . [Animal agriculture] should be a major policy focus when dealing with problems of land degradation, climate change and air pollution, water shortage and water pollution, and loss of biodiversity. Livestock's contribution to environmental problems is on a massive scale."[21] Comedian Bill Maher adds, "When it comes to bad for the environment, nothing—literally—compares with eating meat. . . . If you care about the planet, it's actually better to eat a salad in a Hummer than a cheeseburger in a Prius."[22]

Environmental problems only worsen as industrializing countries amass wealth and demand diets with more protein-rich meats. The average Chinese consumer boosted meat intake from 9 pounds in 1961 to 137 pounds in 2013,[23] and larger increases occurred in Thailand, Brazil, and Morocco. Researchers estimate meat consumption will double as the world's population approaches ten billion. Increased meat eating requires increased production of animal feed; soybean exports from North and South America, for instance, nearly doubled from 2008 to 2018, from 73 million tons to 143 million tons.[24]

In the United States, meat has long been the star of every meal. While adherents of the Paleo diet claim our ancestors kept themselves trim on the animals and fish they hunted, the science presents a much more complex picture. Not surprisingly, how meat affects health is both widely examined and hotly contested. A Finnish study conducted over a twenty-two-year period found that participants who ate meat regularly faced a 23 percent higher risk of dying, particularly from cardiovascular disease and colorectal cancer.[25] Yet several articles in the *Annals of Internal Medicine* suggest such hazards are so small that they have little impact on most eaters and that meat provides valuable nutrients and proteins.[26] What is clear is that the average American devours about twice the recommended intake of beef, and nutritionists link excessive consumption of animal proteins to kidney stones, osteoporosis, and cancers.

Eating meat poses other health dangers. University of Minnesota researchers found fecal matter in 69 percent of pork packages and 92 percent of poultry; other studies discovered poop in every sample of ground beef.[27] The Centers for Disease Control and Prevention (CDC) noted that "more [foodborne] deaths were attributed to poultry than to any other commodity."[28]

For years, chefs, dietitians, and scientists have pleaded with consumers to reduce their meat and egg intake, for both health and environmental reasons. Yet only 5 percent of Americans claim to be vegetarian or vegan, and 84 percent of those eaters eventually go back to meat.[29] Per capita meat consumption remains near its all-time high.

If there is to be an overhaul of the American diet, history suggests it will not come from the humble lentil or the soy meat alternatives that have long graced the aisles of health food stores.

Vegans and vegetarians for the past twenty to thirty years have enjoyed access to plant-based alternatives such as Boca Burger and Tofurkey. Yet sales of such "meat analogues" remain limited, and meat eaters complain about their flavor. Today's innovators, in contrast, use science to develop meat and egg alternatives that appeal to broader audiences, who judge them to possess improved texture and to taste more like animal-based beef, chicken, and pork.

Some critics of Big Meat call for a return to preindustrial days, when animals roamed freely on the range and ate grasses rather than being packed into crowded feedlots and fattened with grains and antibiotics. These reformers want livestock to move regularly to new pastures, where they eat the tips of grass but allow roots to expand, thereby preserving erosion-curtailing ground cover. Herded to fresh grasslands, farm animals enhance the soil with their organic wastes,[30] and grass-fed creatures reduce the need for the synthetic fertilizers and pesticides used to grow corn and other grains for Big Meat's stockyards.

Yet, as often happens with agriculture, environmental trade-offs balance these benefits. Since grass-fed cattle belch up more methane—and since they spend more months burping because they take longer to reach their slaughter weight—they emit about 20 percent more greenhouse gases than do their grain-fed counterparts. Such criticisms are not meant to reject free-range practices, but they should encourage an openness to innovative approaches being advanced by high-tech entrepreneurs.

Big Meat, worried about competition and criticism, hires doctors to exhort animal-based meat's health benefits, scientists to extol how advanced feeds and antibiotics increase average cattle weight, and sustainability officers to suggest cattle provide manure that nourishes the soil. Feeling misunderstood, the industry pays public relations consultants to promise ambitious reductions in greenhouse-gas emissions. While those image makers admit that killing animals in massive slaughterhouses is not pretty, they try to paint a quaint image of small farmers feeding and caring for free-roaming cattle, pigs, and chickens.

Big Ag also employs high-priced lobbyists to maintain (or expand) its own government subsidies as well as to advance legislation and regulations that discourage sales of plant- and cell-based meats, labeling these modern alternatives "fake meat" or "imitation meat." (One alternative-meat innovator retorts that cattlemen should label their products as "Processed in a slaughterhouse" and "Contains aerosolized fecal bacteria!")[31]

As explained in this part's six chapters, alternative-protein entrepreneurs are overcoming numerous challenges, not the least of which is to demonstrate their progress toward sustainability and equity. With significant transparency, they must devise and continuously refine products that solve environmental and health problems. They must convince profit-seeking funders to invest. They must grow their demand, scale their operations to meet it, and outmaneuver their competition.

To appreciate how innovators can feed the world, cut pollution, and avoid animal slaughter, meet six disruptors who are successfully making tasty proteins from stem cells, plants, insects, and algae.

CHAPTER 1

Josh Tetrick, Eat Just—
Rethinking the Chicken and the Egg

Josh Tetrick's early life did not hint that he would become hell-bent on destroying the conventional chicken and cattle industries. Born in Birmingham, Alabama, he claims to have enjoyed "a meat-and-potatoes childhood," and at other times he asserts he "was pretty much fed on a steady diet of cinnamon rolls out of vending machines, nachos and cheese from 7-Eleven, Burger King chicken sandwiches, and pretty crappy cafeteria food." He says he grew up like most Americans, eating "in a way that is not the best for your body, certainly not the best for the planet."[1]

Tetrick played linebacker on West Virginia University's football team before heading to Cornell University and then to the University of Michigan Law School. It was there that doctors diagnosed Tetrick with hypertrophic cardiomyopathy, a heart condition that prompted him to abandon sports and become a vegetarian; more importantly, it gave him an appreciation for "life's fragility" and sparked a full-speed-ahead cockiness. Two years after graduating, in 2010, that health defect caused a near-death experience, prompting Tetrick to ask more fervently, "What would I do with my life if I knew I only had five years to live?"[2] In fact, his phone every morning at 8:30 a.m. flashes the message: "Prepare to die today," reminding him to make his time count.

While on a Fulbright fellowship, the blue-eyed, almost-six-feet-tall Tetrick turned to philanthropic work in developing countries. He began by helping Liberia's government reform its investment laws, something he claims to have been totally unqualified for, and then he taught street kids to read and transitioned child prostitutes into schools. Yet it was in Africa that Tetrick first embraced the challenge, which became an obsession, of providing nutritious food to people at all income levels. He "saw a lot of kids struggling, saw a lot of kids hungry, saw a lot of kids not eating well," he says. "Even though they were eating differently than I was eating in the South, I certainly saw the connection to them, and that we both were growing up not eating very well."[3]

Of his humanitarian work in Africa, Tetrick says, "It was all things that on paper felt nice. But for me, the experience with nonprofits, the experience with international institutions wasn't doing it."[4]

Upon returning to the United States, he worked briefly for a law firm, for TOMS Shoes, and for 33needs, a crowdsourcing website for social start-ups. The law job lasted a few months, until Tetrick wrote an op-ed that was published in the *Richmond Times-Dispatch* criticizing factory farms, even though his firm, McGuireWoods, represented Smithfield Foods, the world's largest pork processor. He then tried his hand at motivational speaking, imploring high school seniors to embrace social entrepreneurship.

Tetrick had held no job longer than one year, worked out of a friend's studio apartment in Los Angeles, and possessed no background in science, nutrition, or business management. Yet in 2011, at the age of twenty-nine, he and Josh Balk launched Eat Just, initially called Hampton Creek, to create "healthier, more affordable and more sustainable foods."[5] Their goal was to develop alternatives to conventional animal-based products—ones people would actually eat.

In his slight Southern drawl, Tetrick describes Eat Just as a "tech company that happens to be working with food."[6] He chose the revised name to suggest the corporation's virtue.

Friends refer to Tetrick as a natural-born salesman. Probably using crasser terms than at his motivational-speaking engagements, the new CEO outlined his first fundraising pitch to a venture capital firm: "Food's fucked up, man. Here's why. Here's an example. Here's what we're thinking about doing."[7] The brashness worked, and he received half a million dollars from Khosla Ventures, one of Silicon Valley's most sophisticated investors. Although Tetrick refuses to discuss Eat Just's finances, published reports say the company in March 2021 surpassed $650 million in investment, and business analysts label the firm a "unicorn," one of the rare start-ups valued at more than $1 billion. (By that date, Eat Just's production had also surpassed the equivalent of one hundred million eggs.)[8] The entrepreneur has discussed going public when Eat Just reaches operating profitability, which he expects to achieve by the end of 2021.

The current food system, Tetrick argues repeatedly, "is failing most people in the world." Approximately 1.1 billion people "go to bed hungry every night," another 6.5 billion are "just eating crappy food," and about 2.1 billion from both groups are "being fucked right now" by nutrient deficiencies.[9] The CEO acknowledges the importance of carbohydrates, fats, and vitamins to a nutritious diet, but he focuses on proteins because they contain the nine amino acids essential for human health.

Conventional protein production through animal meats, Tetrick continues, is unsustainable. It confines millions of cattle, pigs, and chickens into cramped spaces, where their pathogen-loaded poop leaches into rivers and lakes and the cattle's methane-laden belches accelerate climate change. Farmers plow vast swaths of land—and spread enormous quantities of chemicals—to grow the crops that feed those animals, and then

Josh Tetrick, CEO of Eat Just. *Credit: Nick Klein.*

we slaughter them. As the world's population grows and incomes rise, meat production may double by 2050 to 455 million pounds annually. Says Tetrick, "Our planet cannot afford to supply the water, fuel, pesticides, and fertilizer that industrialized animal production requires. It can't afford the polluted water or the biodiversity loss. It can't afford the moral inconsistencies."[10]

Tetrick also points to a 2020 United Nations report that links animal agriculture to zoonotic diseases, including those caused by the COVID-19, Ebola, and West Nile viruses. He argues that reducing the global demand for meats, which the study says increased by 260 percent in the past half century, is the best means to break the chain of transmission.[11] Trying to find a bright side to the COVID-19 pandemic, Tetrick suggests that it has encouraged consumers and producers to seek and deliver healthier foods.

Tetrick decided initially to examine the properties of plants, noting that more than three hundred thousand species around the planet had not been analyzed, in order to find environmentally sound alternatives to animal-based proteins. Believing plants hold the key to making cookies, pasta, ice cream, butter, and scrambled eggs better, Tetrick and his team began targeting species that would be healthy, sustainable, and tasty. What started as individualized analyses quickly became an automated discovery platform.

The young firm's first plant-based product was egg-free mayonnaise, made mostly from yellow split peas grown in Canada and northern US states. Demonstrating Big Ag's fear of competition, Unilever, which makes and markets Hellmann's mayonnaise, threatened to sue the start-up, claiming that a World War II–vintage regulation defined "mayonnaise" to include eggs. Spurred by an outraged Tetrick, the media printed David-versus-Goliath analogies that mocked the food giant, which quickly backed off; yet to avoid future lawsuits, Tetrick added "egg-free" to his Just Mayo label.

After producing egg-free cookie dough and chocolate chip cookies, Eat Just focused on finding an alternative to the egg itself, which Tetrick describes as "the most common protein on the planet."[12] Chickens lay 1.4 trillion (with a "t") eggs each year, but the industrialized layer industry pollutes the air with methane and the waterways with nitrates and harmful bacteria. Hens live in cramped cages and eat so much of a weight-gaining mixture that their legs often collapse. Tetrick claims he's out "to do it better" and that he's "building and fighting for a healthier food system."[13]

Finding a plant-based substitute required a good bit of trial and error. "We tried scrambling the yellow split pea, which kind of just evaporated in the pan," Tetrick says. "We tried grains like sorghum, which tasted like eating wheat bread, and we tried another grain that gelled in the pan nicely, but tasted like tree bark."[14] He finally settled on the mung bean, a legume grown mostly in Asia that gels and scrambles like an egg when ground into a liquid; to achieve the light yellow of a pan-cooked egg, he added turmeric. Eat Just offers "ready to scramble" liquid eggs as well as frozen products ready to pop into skillets, toasters, microwave ovens, or conventional ovens.

Peet's Coffee, Philz Coffee, and other large restaurant outlets agreed to sell Eat Just's liquid eggs at a premium, targeted to vegans and eaters wanting to avoid cholesterol. Yet to expand his market, Tetrick knows he must drop his costs by about one-third and reach price parity with conventional eggs, approximately five cents for a 1.5-ounce serving of protein. To cut expenses, he plans to source mung beans from more regions and to buy them under contracts rather than on spot markets. To ensure reliable supplies, he sets quality standards for bean growers. To increase his yields, he expects dramatic improvements in Eat Just's protein extraction process, and to gain additional revenue, he sells his leftover starch and fiber to the Emsland Group, a giant food-processing company.

To increase his production, Tetrick in 2019 acquired a 30,000-square-foot facility in western Minnesota that operates 24/7. A year later, he partnered with European egg giants the Eurovo Group and the PHW Group as well as US-based Michael Foods, a subsidiary of Post Holdings and one of the largest food processors, which each year sells approximately one billion pounds of liquid eggs to school cafeterias, hospitals, and hotels. Eat Just prepares the protein powder, to which these partners add oils and seasonings and then distribute plant-based Just-branded liquid eggs to foodservice retail supermarket chains, including Walmart, Kroger, and Whole Foods Market.

These partnerships are critical, says Tetrick, because Eat Just lacks the capacity to mix, manufacture, and market large quantities of alternative egg products. "We're a food tech company and our focus is identifying functional proteins, converting them into finished products, and building aspirational brands," says Tetrick. "When we have a dollar, we want to invest it in technology; we don't want to focus on things that other companies are better at than we are."[15]

Eggs dominated Tetrick's attention. "The company is Just Egg and Just Egg is the company," he declared in 2019. The global chicken egg market exceeds $238 billion, and he believes humans increasingly will consume egg protein, yet less and less from industrialized layers. "We could be in [a] world in the next ten to fifteen years," he predicts, "where more people are consuming an egg from a plant than from a chicken."[16]

Yet after devising substitutes for eggs, Tetrick confronted the chicken itself, the conventional breeding of which he considers inhumane and unsustainable.

The basic technology needed to extract a few animal cells and grow them independently has been used for more than a decade by medical researchers creating advanced drugs. Although cost rarely has been an issue for these pharmaceutical engineers, the challenge for Eat Just and its cell-based-food competitors has been to devise a cost-effective medium of liquid nutrients in which meat cells can feed and grow. Put another way, they need to bring cutting-edge science to agriculture.

Initially, Eat Just's growth factors came from fetal bovine serum extracted from the blood of fetuses cut from pregnant cows. To avoid alienating ethically motivated consumers and to lower costs, Tetrick claims to have developed an alternative plant-based medium, informed by the company's expanding library of plants and their nutrients, but he won't reveal anything about its proprietary contents.

To develop chicken meat without killing an animal, Eat Just scientists picked up a single feather shed by the barnyard's best-looking bird, one sporting a healthy comb and nicknamed Ian. They extracted a cell from that feather as the basic starter material and placed it in Tetrick's liquid mix, where it multiplied and became dense. "The assumption you need to kill an animal to eat an animal," declares Tetrick, "is just a wrong one."[17]

Having promised "to make the best chicken nuggets ever," Tetrick, clad as usual in blue jeans and a dark T-shirt, claims an out-of-body experience during the first tasting, when he and teammates ate cultured chicken chunks while the bright-combed Ian waddled around the table.

Tetrick repeats frequently that Eat Just produces real meat. "If you have a chicken allergy, and I give you our [cell-cultured] chicken breast, you're going to have an allergic outbreak," he says. "Just because you don't have to kill an animal to make [the meal] doesn't mean it's not actually substantively meat."[18]

Eat Just's website claims its chicken nuggets will be available at "price parity for premium chicken you'd enjoy at a restaurant." In December 2020, the firm obtained its first regulatory approval, from the city-state of Singapore, and then made its first commercial sale of cultured meat to 1880, a restaurant on Robertson Quay. Tetrick declared, "This historic step . . . moves us closer to a world where the majority of meat we eat will not require tearing down a single forest, displacing a single animal's habitat or using a single drop of antibiotics."[19] He also stated, "We've been eating meat for thousands of years, and every time we've eaten meat we've had to kill an animal—until now."[20]

After tackling chicken, Eat Just set its sights on beef, but not from any cattle. The company's researchers traveled to Japan, to the foothills of Mount Akagi, where the Toriyama family for thirty years has raised prized animals whose rich Wagyu steaks are renowned for their umami taste and delicate marbling. Considered a delicacy, those meats have been expensive and enjoyed by a limited number. Tetrick claims that "with a few cells from the best cows in the world, we'll be able to bring the Toriyama family's tradition to millions more—building a healthier, stronger and more just food system along the way." Wataru Toriyama puts it slightly differently: "The point is to deliver deliciousness to everyone."[21]

"The interesting thing about cellular agriculture," says Tetrick, "is it's not any more expensive to source a cell from the highest-end cow in the world as the cheapest cow in the world." So while Eat Just plans to first market its cultured Wagyu meat (as well as Iberico pork) to Michelin-starred restaurants and premier supermarkets, "we want to get the cost of production of those below even the production of the cheapest meats, and that's how we end up really shifting the system."[22]

In classic Northern California high-tech style, Eat Just's first head-quarters was the garage of a modest home, on Tenth Street in San Francisco. The company now occupies an expansive, 98,000-square-foot former chocolate factory in the city's Mission District. In one of the warehouse-like rooms, scientists isolate cells plucked harmlessly from animals, while those in another lab nourish the cells in large flasks that machines shake gently to encourage rapid growth. Other stations convert the best-performing tissues into food products.

Tetrick admits to an "explicitly aggressive" approach to running his company. Noting his heart condition and the uncertainty of his time on the planet, he declares his intention to save the world and feed a growing population. "I'm fucking going for it, man. I'm fucking going for it," he states. "Life is short. This mission, there is no separation between the company and the mission for me. I am going to do everything I can and use my life and team here to change things."[23]

That unapologetic style has ruffled feathers, so to speak, even within his company. In August 2015, Business Insider reported that former staffers questioned the firm's employment practices, and Bloomberg the following year suggested in a series of articles that Eat Just unethically bought mass quantities of its own mayonnaise in order to inflate perceptions of the product's popularity. In June 2019, three senior executives concocted a "coup" against Tetrick, who hired a security firm to track their revealing emails and then sacked them. Other staffers quit and accused the company of exaggerating profit projections. Target, one of Eat Just's biggest outlets, expressed concern about salmonella contamination and temporarily pulled the company's products from its shelves. In

July 2019, all four outside board members resigned, saying that the company was moving too aggressively against the meat and egg industries.

Tetrick's bellicose style also attracts attacks from Big Ag. Fearful of growing competition, for instance, the American Egg Board paid bloggers to spread criticism of Eat Just's products. The trade association's internal documents said the tiny start-up represented a "major threat" and "crisis" for the $5.5 billion per year egg industry. A representative of a group member in an email chain asked about Tetrick, "Can we pool our money and put a hit on him?" The organization's executive vice president offered, supposedly jokingly, "to contact some of my old buddies in Brooklyn to pay Mr. Tetrick a visit."[24]

Tetrick's critics grow as Eat Just expands. Cattlemen and broiler processors lobby state legislatures to label his new products as "fake meat" or "lab-grown meat." Rural politicians fear meat alternatives threaten farm-based economies tied to traditionally raised chickens, cattle, and pigs. A few agricultural academics complain that cell-based proponents denigrate beef cattle that can graze on marginal lands unsuitable for cultivated agriculture.

Tetrick admits that to temper his opponents and attract more allies, he must provide more economic and environmental data. Noting that culturing meat requires lots of energy, he needs to demonstrate how efficient his process can become and how it will compare favorably with the conventional feeding and processing of farm animals. He also must show how quickly Eat Just can scale up production, whether cultured meats will be affordable, and how consumers will react to food produced differently.

Since Tetrick does not think consumers will revolutionize their own eating habits, he plans to revolutionize the food industry. "Our theory is that given that human beings have been eating meat for about 2.4 million years, it's a hard sell to get them to stop eating meat now, especially now that most of humanity is rising up out of poverty," says the young CEO. "The best way to deal with the meat challenge is just to make better meat without all the issues associated with killing animals today."[25] Referring to his products as "clean meats," he argues that producing plant- or cell-based meats is more sanitary—leading to fewer recalls or outbreaks of *E. coli*—than killing animals in feces-infested slaughterhouses.

Tetrick claims Eat Just's "no-kill" meat will "change the food system" to benefit consumers and the environment, but he admits it takes time to advance new technologies and transform massive industries.[26] Of course, the hard-charging and ever-confident entrepreneur expects that time will be limited, that his chicken products will be available in small markets by 2021, and that they will be available on a large scale in multiple countries after 2025.

Tetrick knows he must consistently lower costs, increase production capacity, and gain consumer acceptance if he is to shift the world away from industrialized slaughterhouses. Yet he seems to be backing away from direct confrontations with Big Ag and instead forging licensing deals with large corporations—such as Cargill, JBS, and Tyson Foods—that have capacities to build big bioreactors and market clean meats aggressively. Such moves cause skeptics, such as author Jenny Kleeman, to worry that meat substitutes could change little about agriculture because their production might remain controlled by giant corporations reliant on remote technologies.[27]

Tetrick spends more and more of his time in Asia, where he finds intense interest among financiers and consumers in protein alternatives. He increasingly focuses on the near-term profit potential of plant-based egg products, calling his work on cell-cultured meats "a much longer-term endeavor."[28]

For someone whose doctor warns him against strenuous activity, Tetrick appears fit and frenetic, moving rapidly from one lab to another, barking out orders, and gesturing energetically with his hands. Having limited time, he is in a hurry. "We just have to figure out a way to use capitalism [to change how we get our food] in a way that is aligned with the kind of planet we want," he declares, "and that's going to take a lot of work."[29] What gives him hope, Tetrick concludes, "is there's such a fire of entrepreneurship right now."[30]

CHAPTER 2

Uma Valeti, UPSIDE Foods— Avoiding Animal Slaughter

Uma Valeti took a different route to cellular agriculture. Born in Vijay-awada, India, Valeti first started thinking about animal welfare at a neighbor's twelfth birthday party. In the front yard, children devoured chicken tandoori and curried goat, while in the back, cooks decapitated and gutted animals. "It was like, birthday, death day," he recalls. "It didn't make sense."[1]

Valeti subsequently read about the inefficiencies of conventional meat production, but "what bothered me more than the wastefulness was the sheer scale of suffering of the animals," he says. "It pained me so much to see them at the markets, lined up and marched to their death, which it seemed to me they knew very well was coming."[2] He laments that humans kill two hundred million animals each and every day for food.

Although still a part-time carnivore when he arrived in the United States for his cardiology residency at the Mayo Clinic, Valeti became increasingly concerned about foodborne illnesses from contaminated slaughterhouses. "I loved eating meat," he said, "but I didn't like the way it was being produced. I thought, there has to be a better way."[3]

The physician began to think about alternative meat production when studying how stem cells repair muscle tissue damaged by heart attacks. "I came to the realization," Valeti says, "that if stem cells could be used to regrow heart muscle tissue, they should be able to be used to grow animal muscle tissue—aka meat."[4] "And while you're at it," he states, "why not grow the steak with a healthier nutritional profile?"[5]

Valeti quickly realized he needed to overcome several obstacles if he were to manipulate stem cells—as well as muscle and connective tissue cells—into a chicken nugget or T-bone steak. Cultured cells, for instance, do not automatically assemble themselves into familiar food shapes, and thicker cell-grown tissues lack the blood vessels and veins needed to carry nutrients to inner muscle tissue. The researcher, however, is encouraging these cells to grow around customizable nanofiber edible scaffolds, similar to the way trellises support the growth of grapevines; each scaffold within the liquid growth medium represents a tenderloin or another cut of meat. One scaffolding option is a mycelium, the vegetative part of a fungus, such as a mushroom, that features a network of fine filaments.

Valeti tried to convince others to start such a business, but several entrepreneurs expressed skepticism. So he decided to go out on his own, hooking up with Nicholas Genovese, a stem cell biologist, and Will Clem, a tissue engineer. "I came to the realization," he says, "that if I continued as a cardiologist, I might save maybe a few thousand lives over the next thirty years. But if I am successful in helping to change the way meat is made, I could positively impact billions of human lives and trillions of animal lives."[6] So this Mayo Clinic–trained heart doctor decided in 2015 to develop a healthy alternative to beef and founded UPSIDE Foods, headquartered in Berkeley, California.

The academic in Valeti initially led him to call his firm Crevi Foods, building off the Latin word for "origin" or "to arise." Yet he soon admitted, "Nobody understood the name." In searching for an alternative, he turned to Clem, who, somewhat ironically, is a certified barbecue

Uma Valeti, CEO of UPSIDE Foods. *Credit: UPSIDE Foods.*

pitmaster whose family owns a chain of meat restaurants in Memphis, Tennessee. In a nod to Clem's hometown, as well as to the ancient city that served as Egypt's capital, they settled on Memphis Meats, but, to signify the commercial launch of their chicken product in May 2021 they changed the name again to UPSIDE Foods.

Valeti—who, unlike Josh Tetrick, has maintained a singular focus on cultured meats—rattles off numerous reasons why conventional animal agriculture must change. Feeding livestock demands about one-third of the planet's arable land and fresh water. Belching cattle emit far more greenhouse gases than all transportation vehicles combined.

The factorylike feedlots that fatten pigs and chickens emit tremendous amounts of toxic excrement, and they tend to be breeding grounds for disease; according to one study, nearly half of supermarket chickens are tainted by feces, based on the presence of E. coli bacteria.[7] As noted before, conventional animal agriculture also wastes calories, since only a fraction of the feed's energy becomes part of the meat we consume.[8]

Yet Valeti's major grievance against industrial animal agriculture is its cruelty. Even beyond the mass slaughters, he complains about the animals' living conditions; most egg-laying hens, for instance, cannot walk about or spread their wings in cages measuring sixty-seven square inches, the floor being a bit smaller than a piece of printer paper.

In a nod to Peter Singer, the Princeton University ethicist and author of *Animal Liberation*, Valeti suggests that clean-meat producers are advancing the centuries-old "rights revolution" that moves us away from all forms of oppression.[9] Although a critic of industrialized agriculture and its abuse of animal welfare, Singer, like Valeti, embraces technology. "The general attitude that technology is a bad thing is a mistake," the professor says. He sees innovative machinery as "an extension of our humanity rather than a loss of it."[10]

Despite these concerns with our conventional protein-delivery system, beef, particularly the hamburger, has become an American icon whose popularity spreads around the world. US consumers each year devour fifty billion burgers, or about three per week for everyone living in this country. "Our target consumer," says Valeti, "is the meat lover who doesn't want to compromise."[11]

Fabricating a conventional hamburger requires a fair amount of science and engineering. Big Meat's researchers spent years examining how the human mouth grinds and blends food, and, in order to leave a pleasant sensation, they manipulate the patty's mince to be neither too coarse nor too fine. They also design slaughterhouses to be assembly-line marvels.

Valeti argues that cultured meat expands significantly on that science and engineering and does it under more controlled conditions, with no slaughter, and with less chance of microbial contamination. Admitting that the process is complicated, UPSIDE Foods' technicians extract cells painlessly from an animal's muscle tissue, as well as from its fat and connective tissue—including ligaments and fascia—that enhance the flavor of meat. They insert those cells into a growth potion that includes nutrients, salts, and acidic buffers. The company avoids the use of fetal bovine serum, which Valeti says costs too much, produces inconsistent quality, and is derived from the killing of animals, which runs counter to his start-up's anticruelty mission. The cells multiply in bioreactors, or what Valeti calls "cultivators," that look and act a lot like fermentation tanks within a beer brewery. To keep the cells growing, the bioreactors maintain a constant temperature, regulate pH levels, and manage nutrient concentrations.

Viewed through a microscope, Valeti's cultured cells wriggle, forcing a reconsideration of what it means to be alive and what defines something as animal flesh. When asked how the multicellular chains flex without a brain and nerves, a company technician explains, "This is all living tissue. . . . For the past 12,000 years, we've assumed that when I say the word 'meat' you think 'animal.' Those two ideas are concatenated. We've had to decouple them."[12]

UPSIDE Foods had developed enough lab meat by January 2016 to make its first tiny meatball. "This is absolutely the future of meat," declared Valeti. "We plan to do to animal agriculture what the car did to the horse and buggy. Cultured meat will completely replace the status quo and make raising animals to eat them simply unthinkable."[13] A little over a year later, the start-up created its first "clean" fried chicken and duck à l'orange.

Valeti hopes to build a large production facility by 2022, but he will not commit to a date for his first commercial sales, saying only that they will be in "the near future."[14] His initial target audience will be consumers viewing clean meat as something worth paying a bit more for, as shoppers already do for wild-caught salmon or grass-fed sirloin. He also intends to market cultured meat for its positive impact on climate change; professors at the University of Oxford and the University of Amsterdam asserted that, compared with industrial agriculture, cell-based meats cut greenhouse-gas emissions by up to 96 percent.[15]

Valeti admits some consumers cannot stomach the idea of eating beef, chicken, and pork produced in a bioreactor. A 2018 opinion poll found that only 20 percent of Americans were likely to buy "lab-grown" meat,[16] but Valeti challenges the polling question, countering that his meats, like most products in a supermarket, will be produced in a factory rather than a lab. He references a comparison made by journalist Mary Catherine O'Connor: "If given a choice between, on the one hand, holding a pneumatic gun up to a 1,200-pound cow, right between its eyes, and shooting a steel bolt through its skull to render it unconscious before it is eviscerated, or using a syringe to take a biopsy from a live cow and then bringing that sample to a bioreactor where its muscle stem cells are engineered into meat . . . well, which process is more icky?"[17]

Valeti acknowledges that cell-cultured meat "only three years ago would have been considered science fiction," and he recognizes that some consumers will think it "weird" to eat. Yet the entrepreneur argues people will embrace such offerings once they experience their meaty tastes and understand their environmental and health benefits. He further maintains that as consumers care increasingly where their meals come from, they will welcome food choices that do not involve the slaughtering of animals.[18]

A few animal-alternative advocates, responding to the "yuck factor," focus on fashion goods, arguing that crafting cell-based leather is less complex than growing meat in a lab. They say the $100 billion leather industry is ripe for disruption because it deploys powerful chemicals to tan, or mummify, belts and watchbands so the cow's hide does not rot away. Finally, they argue consumers will more readily adopt a lab-based product they wear than one they eat; getting shoppers comfortable with cultured leather, these advocates suggest, will make customers more willing to try cultured meats.

Valeti disagrees and wants to concentrate on food. He expects to engage influential chefs and independent researchers to proclaim his meat's tastiness as well as its nutritional and environmental benefits. Yet he argues that the cultured-meat industry, if it is to penetrate markets, must launch open and honest conversations with consumers: "This has got to be a production process that's a lot more traceable than the current industry."[19]

Politicians and bureaucrats also battle for a role regulating this emerging industry. In the United States, oversight of new production processes prompted clashes between the US Department of Agriculture, which monitors meat and eggs, and the US Food and Drug Administration, which regulates biologics, including products derived from tissue cultures. After several years of negotiations, regulators from both agencies, using bureaucracy-speak, reached an agreement in March 2019 "to foster these innovative food products and maintain the highest standards of public health."[20]

Competition for supplying alternative protein is fierce, and different entrepreneurs deploy different descriptors—"slaughter-free," "cultivated," "clean"—and try diverse technical and marketing approaches. Israel-based Aleph Farms, for instance, 3D prints meat on the International Space Station; Wild Earth, based in Berkeley, California, grows cultured mouse meat for cats; and Mission Barns creates duck sausages. A few start-ups, such as Massachusetts-based Galy, produce nonfood products such as cotton, developed from the plant's cells, to provide yarn more sustainably for clothing. At least sixty cell-based meat firms operate around the world, and they attracted $314 million in venture capital in 2020.[21] One market analyst projects their global sales will rise from $214 million in 2025 to $593 million in 2032,[22] while a separate business researcher estimates that 35 percent of all meat will be cultured by 2040.[23]

Big Meat worries about competition from alternative meats, but a few corporate giants do not want to be left out of this promising market. An executive at Cargill Protein, which invested in Aleph Farms and UPSIDE Foods, asserts, "Cultured meats and conventionally produced meats will both play a role in meeting [consumer] demand."[24] The CEO of Tyson Foods, which bought a minority stake in UPSIDE Foods, adds, "People want protein, so whether it's animal-based protein or plant-based protein, they have an appetite for it."[25] Valeti welcomes the involvement of Big Meat, advocating a "Big Tent philosophy where meat eaters and meat producers across the world see [cultured meat] as a solution that everyone can get behind."[26]

The entrepreneur in Valeti, however, wants to maintain corporate control, and he recognizes that his costs must fall while his production capacity rises. To defend his optimism about both, he points proudly to UPSIDE Foods' consistent improvements in cultivating and duplicating cells. His meat-production process, which used to take several months, now occurs in three weeks, much less time than is needed to raise an animal to slaughter weight. While UPSIDE Foods won't divulge its cost projections, Israel-based Future Meat Technologies claims its cultured beef will by 2022 be priced as little as $10 per pound—still about 20 percent higher than grass-fed meat but quite a drop from the $1.2 million per pound paid by Mark Post and Maastricht University in 2013. "Our goal is to get to cost parity," Valeti says, "and then beat commercial meat."[27]

The CEO of Whole Foods Market shares UPSIDE Foods' optimism. "The environmental impact of the meat industry cannot be underestimated," says John Mackey. "It is very challenging and hence the need for us to innovate towards these cellular-based technologies." He predicts such alternatives will transform the meat industry in the next fifteen years, and he adds, "I think they're going to be able to do it cheaper" and without polluting lands or releasing greenhouse gases.[28]

The scientist in Valeti admits there is much still to be discovered. We don't know, for instance, how important an animal's muscle movements are to the structure and taste of its tissues. We don't understand whether the lack of an immune system impacts cultured meat. We don't appreciate why cells from certain species exhibit genetic changes after several duplications.

Valeti the visionary wants to transform cattle, chicken, and pig farming. "The status quo in animal agriculture is not OK," he asserts. "That status quo is going to kill a lot of people."[29] His mantras include "Think of our operation as a farm at a tiny scale," "Make meat without slaughter or disease," and "Sustainably feed the world." Such goals have attracted funds from Bill Gates, Richard Branson, and other high-profile investors;[30] by mid-2020, the firm had raised more than $180 million.

Opinions vary about Valeti's prospects and the speed and extent of his food transformation. RethinkX, an independent think tank, predicts alternative proteins will make industrial animal farming obsolete, and it foresees that "by 2030, the number of cows in the U.S. will have fallen by 50% and the cattle farming industry will be all but bankrupt."[31] Ermias Kebreab, an animal agriculture expert at the University of California, Davis, disagrees, saying, "I'd rather have beef" from cows, and he favors neo-agrarian farming—with cover crops and manure—over cellular agriculture.[32] A more moderate tone comes from Isha Datar, executive director of New Harvest, a nonprofit group promoting research on cultured meat; she declares that cell-based products have enormous potential but are not a certainty to displace industrial agriculture.[33]

As part of the pending food transformation, UPSIDE Foods' cofounder Will Clem envisions the company's technology altering eateries. "A lot of restaurants have a beer tank in the corner, and they're brewing an IPA," he observes. "Well, this is the same thing, only it's growing beef, pork, or chicken."[34]

Valeti thinks globally, proposing to produce diverse meat products in response to different cultures and consumer demands. Since chicken is America's most popular meat, he unveiled in 2018 lab-cultured nuggets. Next up was "clean" duck, which he hopes will be particularly popular in China and France.

The entrepreneur also thinks historically. "Humans evolved in large part because we domesticated livestock," Valeti asserts. "If we can produce meat without slaughtering animals, we'll have launched a second domestication."[35]

CHAPTER 3

Patrick Brown, Impossible Foods—Making Burgers from Plants

Patrick Brown's path to his new passion began with a medical degree and a three-year pediatric residency at Chicago's Children's Memorial Hospital. Yet, like Uma Valeti, Brown decided against treating individual patients, feeling he could make a bigger impact by researching how to block particles, such as the AIDS virus, from infecting other cells. He became a biochemistry professor at Stanford University and developed DNA microarrays that scientists now use to map and monitor gene activity, as well as to distinguish between cancerous and normal tissues; that discovery has been described as "one of the most important genomic advances of the decade."[1] Brown also cofounded the Public Library of Science, which provides open access to published scientific research, and among his many honors is the National Academy of Sciences' Award in Molecular Biology.

In 2009, however, Brown took an eighteen-month sabbatical to consider what greater challenges to confront, and he decided that the biggest global problems result from how we raise and slaughter animals to supply dietary protein. He initially wrote academic papers and organized workshops on the topic, but he concluded that the most effective action would be to abandon his Stanford professorship, which he claims to have been his "dream job," and develop an animal-free product that competes in the free market. He founded Impossible Foods in July 2011.

Brown, a trim marathon runner, headquartered his start-up in a low-rise office park in Redwood City, between San Francisco International Airport and Stanford University. He declares, "The use of animals in food production is by far the most destructive technology on earth."[2] Brown applauds the growing public protests against animal slaughter, such as Joaquin Phoenix's attack on such "cruelty" during his 2020 Oscar acceptance speech. Yet convinced he could not use words and statistics to guilt-trip consumers into eating less meat, Brown decided the only way to change eating patterns was to create an indistinguishable substitute.

Although a devoted vegan, Brown appreciates that many meat eaters will not give up their carnivorous habits readily. He recognizes that numerous consumers have enjoyed the taste of steak, chicken, or pork since they were children, and he's heard the arguments that violence associated with animal slaughter is ancient and perhaps part of a deep human desire to exert control over our surroundings. Such observations, however, do not stop him from promoting plant-based meat substitutes, even to those hard to convince.

"The only consumer we care about is the hardcore meat lover," declares Impossible's CEO. In fact, 90 percent of the company's customers are meat eaters. "I love vegetarians and vegans as much as the next guy, but that is not the customer we care about."[3] Part of this is smart marketing. Few consumers claim to be vegetarians or vegans, so Brown stresses that his Impossible Burger looks and smells like a beef patty. Sarah Schafer, the chef-owner of Irving Street Kitchen, an upscale meat-focused restaurant in Portland, Oregon, agrees. "A lot of vegans and vegetarians don't want anything that's going to imitate meat in any way," she says. "It's more the meat-eaters who order it. They're like, 'Oh my God, this is just like eating a burger.' It has an iron-y flavor to it."[4]

Patrick Brown, CEO of Impossible Foods. *Credit: Impossible Foods.*

Brown says he had a hunch that heme, the iron-rich oxygen-carrying compound in blood and muscle tissue, is what gives meat both its protein and its unique flavor. He recognized that plant-based sources of heme are numerous and that the challenge is scalability, so he initially focused on widely available clover and soy roots. "I dissected them with a razor blade and then blended the shavings to see what I could extract," Brown explains. He spent five years painstakingly trying numerous combinations, settling finally on heme from the fermented roots of soybeans. To devise a plant-based burger, he mixed it with potato protein, sunflower and coconut oil, yeast extract, salt, vitamins, and several flavorings. The result offers a bloody-red-meat-like appearance that turns a grayish brown when cooked. The firm's flavor scientists conduct one hundred assessments each week as they evaluate alternative spices and seasonings.

In addition to designing a product that looks and tastes like ground beef, Brown and his colleagues devised a mixture that feels familiar in our mouths, which is not easy because what happens when we chew meat is rather complex. Initially, our teeth split apart the bundles of muscle fibers before the enzymes within the mouth reassemble them into the viscous blob we swallow. Obtaining the same effect with plant-based fibers required more experimentation.

The combination of mouthfeel, appearance, and taste has created a growing market for Brown's patties. By mid-2021, Impossible Burgers were being sold at more than twenty thousand grocery stores and thirty thousand restaurants, including the Burger King and White Castle chains.[5] At Whole Foods Market and Kroger, sales rose by 23 percent when the supermarkets moved plant-based meats from their own isolated section to be next to animal-based meats, where shoppers welcome head-to-head comparisons. In June 2020, Impossible Foods launched an e-commerce site that allows free two-day delivery of its products throughout the lower forty-eight states. Although still a niche market compared with the $176 billion beef industry, consumer orders are rising so rapidly that Impossible Foods struggles to increase production,[6] and the company in late 2020 announced expansion across Asia and Canada.

To meet growing demand, Brown's plant-based meat company in September 2017 opened its own 68,000-square-foot factory in Oakland, California, which makes about 12 million pounds of product each year. Two years later, it partnered with OSI Group, which rents out food production facilities around the world; Brown claims this turn toward dispersed factories increases Impossible's output and ensures the start-up delivers plant-based meats more quickly to customers.

Brown focuses increasingly on industrial fermentation, which uses microbes such as microalgae and mycoprotein to vastly increase the production of protein biomass, including the synthesized heme that allows

Impossible to imitate the flavor and texture of beef. Several large food and life science companies, such as DuPont and Novozymes, also are entering the emerging fermentation sector, and new firms devoted to alternative proteins obtained $274 million in venture capital investments in 2019, up fivefold from the previous year.[7] Impossible's scientists believe they soon will produce clean-meat hybrids made from plants and fermented microbes.

Among fermentation's advantages are speed and efficiency; it takes years to fatten animals and months to grow plants, but microbes can double their biomass in a few hours. This method of producing proteins also uses a fraction of the land, water, and chemicals associated with raising animals or cultivating crops. Yet the infrastructure to manufacture large quantities of fermentation-based products is not remotely here yet.

In the meantime, Brown continues to ramp up his list of products. In January 2020, Impossible Foods introduced plant-based pork that mimics hog meat's supple texture and mild taste; to create sausage, the firm's chefs simply add a mixture of spices. Those new products, says Brown, allow the company to expand internationally because pork is the world's most widely eaten meat and is especially popular in China. "To produce a full range of meats and dairy products for every cultural region in the world,"[8] Impossible also is developing plant-based chicken and melty cheese for pizza.

This level of expansion requires a certain business savvy, and Brown plays the line between academic and corporate rebel. Although he wears suit jackets for special meetings, Brown dons the Silicon Valley uniform of green or purple hoodie when being interviewed on CNBC's business shows. Google early on offered to purchase his start-up, but Brown rejected it "in less than five seconds, because we would have just been one of their suite of nifty projects."[9] He designated forty Big Ag and Big Meat companies that are "disallowed" from making investments in Impossible Foods.

Notwithstanding such limitation, the start-up attracts big-name financiers, including Peter Thiel's Founders Fund, Khosla Ventures, Bill Gates, Jay-Z, Katy Perry, Serena Williams, Horizon Ventures, UBS, and Viking Global Investors. Impossible Foods obtained an additional $200 million in mid-2020, increasing its estimated value to more than $10 billion. Expressing plans to go public in late 2021 with an initial stock offering, Impossible's chief financial officer says, "We're raising enough money to scale up our production and push ahead with our research and development."[10]

Brown maintains that Impossible Foods offers disruptive technologies that will displace agriculture's status quo, and he plans to double his research and development efforts in 2021. "Our products are going to get tastier, healthier, more affordable and better in every way, continuously into the far future," he says, "and the incumbent meat industry is just standing there, waiting for the tsunami."[11] He points to buggy-whip manufacturers as examples of staid corporations that couldn't compete with new approaches. "The horse was just as deeply embedded in the culture as the cow is today," he observes. "It took a couple of decades before it was obsolete. Better technology comes along and it's a new game."[12]

Brown believes his better technology produces foods that customers prefer and that, in the process, reduce pollution and animal cruelty. He boasts that making meats from plants emits 89 percent less greenhouse gases than making a burger from cattle. The innovative process also uses 92 percent less land and 74 percent less water. "If we succeed in our environmental mission," he says, "we will be the biggest, most impactful business in history."[13]

Not all agricultural reformers and environmentalists see it that way. Advocates of regenerative grazing accuse Brown of trying to bankrupt small farmers, ranchers, and rural economies that depend on raising animals for food. Friends of the Earth complains that the Impossible

Burger "implicates the extreme genetic engineering field of synthetic biology, particularly the new high-tech investor trend of 'vat-itarian' foods."[14] People for the Ethical Treatment of Animals (PETA), which you'd think would welcome Brown's efforts to slow down the slaughter of cows, pigs, and chickens, attacks his company for having tested alternative ingredients on 188 rats. Brown, a vegan who used to support PETA, responds to the group's criticism: "With a lot of fundamentalist religious groups, it's bad if you're a nonbeliever. But if you're a *heretic*—that's a capital crime."[15]

While Brown vigorously defends his company's environmental bona fides, he does admit the Impossible Burger is not the most nutritious food, just a more nutritious hamburger. "The niche that this fills is not the same niche that a kale salad fills," he says. "If you're hungry for a burger and you want something that's better for you and better for the planet that delivers everything you want from a burger, then this is a great product. But if you're hungry for a salad, eat a salad."[16] He boasts that the Impossible Burger includes no antibiotics or cholesterol but does offer a rich mix of proteins, iron, and nutrients.

Critics challenge Brown's health claims. "Let's dispense with the idea that this is 'healthier' in any way," says Aaron E. Carroll, a pediatrics professor at the Indiana University School of Medicine. "The Impossible Whopper has 630 calories [versus a traditional Whopper's 677]. It also contains similar amounts of saturated fat and protein, and more sodium and carbohydrates. No one should think they're improving their health by making the switch."[17]

The entrepreneur faces numerous plant-based competitors. Beyond Meat, headquartered in California, enjoyed a strong stock market debut in mid-2019, its products sell online and at 26,000 retail locations nationwide, and it projects that by 2024 at least one of its offerings will cost less than animal-based meats.[18] The Sacramento-based Better Meat Co. raised $8 million in seed capital in July 2020 to advance hybrids that blend plant-based proteins into ground meat products. Canada-based Daiya Foods offers, in more than 22,000 stores, nondairy cheeses, a vegan mac-and-cheese substitute, and a pepperoni pizza without meat, dairy, or gluten. Even several agricultural and food conglomerates— including Nestlé, McDonald's, JBS, Cargill, and Kellogg's—are testing or marketing plant-based burgers and nuggets.[19]

Brown also feels competition from cultured-meat producers such as Josh Tetrick and Uma Valeti, yet he believes extracting and growing animal cells is highly inefficient and will not be embraced by consumers. He goes so far as to call cell-based meat "one of the stupidest ideas ever expressed."[20] (Valeti responds, "We will offer real meat, with no compromise. Pat is welcome to come try it!")[21]

That aggressive assertion is common for Brown, whose soft voice and serious smile belie his confrontational style. The CEO displays a plaque on his office wall declaring "Blast ahead!" and asserts that his "favorite thing to do is to get into an argument." His preferred target is Big Meat. To collapse the livestock industry, Impossible decided also to produce plant-based dairy products, thereby "going after every piece of consumer value that comes out of a cow."[22] Brown brashly declares, "We plan to take a double-digit portion of the beef market within five years, and then we can push that industry, which is fragile and has low margins, into a death spiral. Then we can just point to the pork industry and the chicken industry and say 'You're next!' and they'll go bankrupt even faster."[23]

Not surprisingly, Brown confronts growing opposition from Big Meat. The Center for Consumer Freedom, profiled by *60 Minutes* as a front group for food processors and tobacco companies, took out a full-page ad in the *New York Times* in November 2019 to criticize plant-based offerings as "fake meat" or "ultra-processed foods" that can cause weight gain. The Center's website claims the Impossible Burger contains too much salt and compares it to dog food,[24] and it bought a television commercial during the Super Bowl, admittedly airing only in the Washington, DC, market, alleging that plant-based meats contain a chemical laxative called methylcellulose. (Derived from plants, that odd-sounding ingredient actually serves frequently as a binder in sauces and ice cream.) Pat Brown responded with his own ad claiming there's "poop in the ground beef we make from cows," and a company statement added, "Impossible Foods does not tolerate Big Beef's bull."[25]

Impossible upped its marketing in April 2021 with its first national campaign, complete with television, digital, and social elements. Targeting meat eaters who have not yet tried the company's plant-based burger, the advertisements declare, "We are Meat."[26]

Despite rising sales at Burger King and other fast-food chains, plant-based alternatives remain less than 4 percent of the US hamburger market. Sales, however, are growing at an annualized rate of 28 percent. Admitting its estimates may be quite conservative, the Switzerland-based bank UBS predicts that plant-based sales will accelerate from $4.6 billion in 2018 to $85 billion by 2030.[27]

Given that Impossible's sales and investments continue to rise, Brown could become extremely wealthy if the company goes public, but the entrepreneur doesn't want to sacrifice his independence or change his lifestyle. He and his wife still live in the same cedar-shingled condominium for Stanford University faculty that they have occupied for more than thirty years.

Brown claims to be on a mission to disrupt an environment-damaging agricultural system. "We must think outside the box, and certainly outside the feedlot," he says. "We're using innovation to create tastier and more nutritious foods, without harming our planet."[28]

CHAPTER 4

James Corwell, Ocean Hugger Foods— Turning Tomatoes into Tuna

James Corwell introduces himself as a certified master chef, one of only sixty-nine with that title in the United States and one of the few who passed the grueling eight-day test on his first try. He notes that his New Orleans restaurant, Le Foret, was ranked Best New Restaurant in 2010.

In addition to these accolades, the self-confident chef prides himself on turning tomatoes into tuna, or at least imbuing them with the taste and texture of this saltwater delicacy. His inspiration for Ocean Hugger Foods occurred one morning at the Tsukiji fish market in Tokyo, where Corwell witnessed hundreds of dead tuna lying on the floors of two warehouses, each the size of a football field. It was a mind-boggling amount of fish, he recalls. Informed that this single fish auction sold every day a similar volume, about four million pounds, he asked the obvious question: How can the oceans ever keep up? After reading a study in *Nature* that found overfishing had killed 90 percent of tuna and other large fish, he answered his own question: they can't.[1]

According to other research, if trends continue, the planet's seas will support no commercially viable species by the year 2050.[2] Too much fishing disrupts the ocean's entire food web, harming sea turtles, corals, and other vulnerable marine life. It burdens billions of people who rely on seafood as a key source of protein, and it threatens the livelihoods of millions who fish.

Fearing further pressure from a growing human population, Corwell decided to produce fish-free alternatives. Tasty and nutritious seafood, he says, "should be enjoyed by massive amounts of people," but to avoid environmental destruction, we need to "create substitutes that are true to the traditional profile of sushi."[3] The idea of switching tomatoes for tuna began, he says, as "a small idea." He tested it first with a few friends and then at cocktail parties. "People loved the direction of the product," he says, "and they encouraged me to take the product to the next level."[4]

Corwell introduced his first plant-based product, named Ahimi (meaning "the spirit of ahi"), in November 2017 at Whole Foods Market. It contains a mix of five simple ingredients: fresh tomatoes, non-GMO soy sauce, filtered water, sugar, and sesame oil. Tomatoes are the core because they contain high levels of naturally occurring glutamic acids, which also provide meaty foods with their savory flavor. "We start with whole fruits and vegetables," says Corwell. "Then we remove the characteristic flavor from the fruit or vegetable. Finally, we enhance the texture of the produce to mimic that of your seafood favorites and add a simple, flavorful marinade. That's it."[5]

Corwell aims for the meaty flavor of umami, which in Japanese means "a pleasant savory taste"; tomatoes, seaweed, and soy sauce happen to be rich in that taste. He also seeks to construct a texture that "slides across the palate like real tuna."[6] "Food is meant to be eaten in whole," says the chef. "That is what I love about Ahimi and Unami [Corwell's meatless alternative to *unagi*, or freshwater eel]; they are not highly processed. The integrity of the item is left intact. Not everything needs to be highly processed."[7]

Corwell knows eaters will turn to alternative products only if they possess comparable flavors, superior nutrition, or both. To his mind, it's

James Corwell, CEO of Ocean Hugger Foods. *Credit: Ocean Hugger Foods.*

up to chefs to make that happen, and they "need to be leaders in this industry, providing quality cuisine while protecting the environment." He hopes other kitchens will begin to experiment with plant-based proteins and "let customers be the beneficiaries."[8]

Corwell is first and foremost a culinarian, trying to create something that fish eaters will love in dishes such as sashimi, nigiri, poké, tartare, and ceviche. Yet he possesses an activist's urgency, declaring, "Social awareness of global pollution and human impact on the environment really resonates with people more than ever. . . . From my own perspective, the population numbers of 2050 and 2100 are staggering enough to justify the change to start sooner" rather than later.[9]

His intensity is shared by Ocean Hugger Foods' CEO, David Benzaquen, who says the oceans are in crisis, with climate change warming the water and acidifying coral reefs. Pointing also to illegal fishing and overfishing, the executive fears the destruction of aquatic ecosystems, where 90 percent of species live and more than 90 percent of carbon is stored. "With an estimated 50 billion aquatic animals killed for food in the US every year, which is five times as much as all land animals combined," explains Benzaquen, "I want to be a part of saving those species, as well as saving our own species."[10]

The pair are also driven by a human health imperative. Corwell says, "Health and wellness issues relating to lifestyle diseases are a major motivator," and he argues that the start-up company is helping consumers live longer—what he calls "people's core dream." Basing foods on plants, he argues, is "the new thesis, the new gestalt."[11] Noting that traditional cuisines, particularly in southern India, long have relied on plants to offer nutritious and rich meals, he encourages chefs to go back in time but with modern approaches.

Indeed, Corwell and Benzaquen are anything but Luddites. They consider themselves technologists and businesspeople. Benzaquen says, "We also get incredible financial rewards with these efficiencies."[12] Ocean Hugger Foods currently sells to restaurants and supermarket sushi counters, but Corwell hopes to eventually offer Ahimi directly to consumers. And he plans to release other plant-based alternatives to salmon (made mostly from carrots) and freshwater eel (made mostly from eggplant).

In other words, Corwell wants to scale up—without falling into the trap of the conventional food business. Claiming that he "pretty much grew up in a kitchen," the Atlanta-born chef laments that he no longer trusts "how food is produced." He declares, "We are required to view the products we work with as being alive and wholesome. Yet the . . . large majority of the food produced in the U.S. is mass produced, in a practical but quite soulless way, as well as quickly and cheaply to an audience who at the very least wants it cheap."[13]

Recognizing that cost will be a key factor for consumers, New York–based Ocean Hugger Foods formed partnerships with large tomato growers and packagers in Mexico and Thailand. It sells its plant-based tuna at about one-third the price of sushi-grade tuna, and it encourages chefs to cut costs by pairing it with other fish in poké bowls and sushi rolls. Corwell knows that elite restaurants, where he spent most of his career, are accessible to only a small portion of the population, but he hopes Ahimi and other plant-based forms of protein can reach a wider market.

Corwell initially targeted vegans, vegetarians, and flexitarians, and he thought Ahimi would appeal to consumers who suffer from seafood allergies or worry about the safety of raw fish, such as pregnant women, the elderly, and those with compromised immune systems. He then pitched it to people who want to advance sustainability and protect endangered species. As more restaurants, supermarkets, and sushi counters carry plant-based alternatives, he sees himself challenging the foundation of industrialized fishing.

Corwell and Benzaquen believe they can capitalize on a massive opportunity because "the current system is inefficient," and consumers want tasty alternatives that do not involve killing fish. Ocean Hugger, they claim, will grow because it satisfies people who "are craving the pleasure of biting into the flesh of the highest-grade tuna, but they know their wallets, and our planet, can no longer afford it."[14]

The innovators, however, accept that their start-up business is vulnerable, as evidenced by how COVID-19 disrupted the supply chain for sushi. Although Ocean Hugger had been signing contracts with more and more retail outlets, the pandemic prompted consumers to avoid raw fish and even plant-based substitutes that had been sold alongside them. The pair paused operations for several months and relaunched "bigger and better than ever" in early 2021.

Corwell and Benzaquen, like so many entrepreneurs profiled in this book, are convinced they will succeed and that they will locate new outlets—such as college and corporate cafeterias—to sell their products.

Since people are used to eating raw fish prepared for them outside of the home, they place Ahimi in sushi bars rather than on the grocery shelves.

While preferring traditional plant-based cooking, albeit with a contemporary touch, Corwell appreciates that other firms—such as Finless Foods and Shiok Meats—use genetics to grow fish and shellfish from stem cells. Claiming that it is easier to culture fish than meat, Finless produces cell-based bluefin tuna, while Shiok places its shrimp tissue into dumplings and is working on cellular lobster and crab.

Praising the role of all disruptors, Corwell says he became a start-up founder because there is only so much a chef can do in a single restaurant to change eating habits and protect the planet. Benzaquen used to think he would be most impactful as an advocate within a nonprofit organization. "Over time," he says, "I've come to realize that I am most effective as an entrepreneur developing, launching and scaling innovative solutions to fix the system, rather than complaining."[15]

Corwell admits he misses the "hustle and bustle of the kitchen, but the company allows me to effect change on a really big scale." In addition to giving ocean species a break and providing tasty protein options to a growing human population, he says his own transition from restaurateur to entrepreneur "has been a blast."[16]

CHAPTER 5

Virginia Emery, Beta Hatch— Farming Insects

Virginia Emery calls herself the "bug lady." If that designation calls to mind an eccentric older neighbor, the type who wears insect brooches and kooky hats, Emery is anything but. One colleague says of her, "Young, sharp and brilliantly assertive, Virginia Emery embodies a new breed of chief executive in the modern technology marketplace."[1]

After obtaining a doctorate in entomology in 2013 from the University of California, Berkeley, Emery spent several years as a research scientist at a major pest-control company, trying to devise poisons to kill dengue-spreading mosquitoes. Yet she slowly moved from viewing bugs as pests to be killed to seeing them as resources to be utilized.[2] She launched Beta Hatch in Seattle in 2015 to breed high-quality bugs at industrial volumes in order to provide proteins, oils, and nutrients for poultry and fish farms, as well as to deliver organic fertilizer from bug poop, or frass. Insects, Emery claims, "are the most sustainable response to agriculture's challenges."[3]

Farming such creatures is not new, and the United Nations estimates that two million people consume bugs regularly. In many parts of the world, house crickets (*Acheta domesticus*) are eaten directly, dry-roasted, baked, deep-fried, or boiled; they also are grown to feed many species of fish and birds. (Consuming insects is known as "entomophagy," a relatively new term.) We also raise bees to pollinate crops as well as to provide beeswax utilized in balms and candles, propolis deployed as a wood finish, and honey for our own food. We domesticate silkworms to deliver an elastic fiber used in many textiles.

Insect ranching offers many benefits. Bugs are rich in proteins, contain few carbohydrates, and possess amino acids, dietary fiber, and vitamins A and B_{12}. These creatures take up little space, live happily when jammed together, survive without light, breed throughout the year, emit few pollutants or greenhouse gases, and require little feed. Compared with cattle, weight for weight, they release eighty times less methane and consume far less water. They mature and reproduce with amazing speed; crickets, for instance, become fully grown within three weeks, and an individual female lays up to 1,500 eggs. To provide the same amount of protein, cattle need twelve times more feed than crickets, and unlike farm animals, insects rarely transmit diseases such as H1N1 influenza, mad cow disease, or salmonella. To use another comparison, an acre devoted to insects produces five thousand times more protein than an acre planted in soybeans, a typical ingredient of animal feed and a component of two-thirds of all processed foods. "We grow," Emery boasts, "the world's most efficient protein."[4]

Emery creates genetic platforms that enable Beta Hatch to selectively breed insects and customize their nutritional profiles. As a result, the start-up markets individualized feedstocks and fertilizers to meet the specific needs of various chicken breeders and fish farmers.

Emery is partial to mealworm beetles (*Tenebrio molitor*), which grow year-round and eat virtually anything, including contaminated wastes and polystyrene. The mealworm's gut, in fact, is one of the few places where Styrofoam decomposes. "Insects are nature's bio-recyclers," the entomologist declares. "They convert wastes into high-quality protein

Virginia Emery, CEO of Beta Hatch. *Credit: Kurt Schlosser of GeekWire.*

for animals, birds and fish, and their own poop gets turned into high-grade fertilizer."[5]

Beta Hatch's mealworms gorge mostly on food-processing wastes, such as discarded grain and yeast from beer brewing. The company harvests the larvae into feed that contains 59 percent protein and 24 percent fat. Unlike conventional fish meal, the insect-based mixture includes neither heavy metals nor contaminants, and it displaces the scouring and killing of tons of small wild fish, a practice that destroys aquatic ecosystems.

Beta Hatch's automated processes begin at the "ranch," where eggs—about a million of which would fill a beer pitcher—lie in trays stacked vertically in a humidified room. In what Emery describes as "food heaven for mealworms," the insects hatch and begin to eat those food-processing wastes. The Beta Hatch team removes the mealworms' frass (manure), replenishes their food, and then harvests the larvae at two to four months of age.[6] Because Emery and her colleagues control the production process tightly to avoid chemicals and impurities, independent agencies certify the mealworm frass as organic.

Beta Hatch employees refer to themselves as insect entrepreneurs. Since they utilize waste products as feed, they claim to be "industrializing insect agriculture within a regenerative food system."[7] An important aspect of that industrialization is reducing costs. Emery notes that conventional animal feed, mostly corn and soybeans, accounts for 50 percent of the cost of industrialized meat and that 30 percent of total crop production goes to nourish livestock. She views those huge numbers as opportunities, believing her proprietary breeding "enables insects to cost-effectively meet the global demand for animal feed and crop fertilizer."[8]

The insects' frass, moreover, contains a mix of plant-nourishing nitrogen, phosphorus, and potassium. For indoor plant use, the compost emits no odors, contains no insects or insect eggs, and can be added dry to small pots. Outdoors, the chitin-rich product defends the soil against nematodes and pathogens while replenishing essential nutrients.

No doubt Beta Hatch is tiny compared with the giant feedstock corporations, yet its advanced genetics has allowed it to expand quickly and become recognized. Emery's facilities in a few years increased production to more than a ton each day of insect-derived protein-rich animal feed, and the CEO published the first scientific paper on the complete genome of yellow mealworms. To make her products more attractive environmentally, Emery uses waste heat from data farms to keep the worms warm and vibrant. To increase her operation's efficiency, she continues to study insect biology, experiment with rearing conditions and diet formulas, and push mechanization throughout her process. As Emery puts it, "We're automating insects for agriculture."[9]

Beta Hatch's progress attracts investors. Leading its 2020 raise were Innova, an early-stage venture capital firm, and Wilbur-Ellis, a $3 billion marketer of agricultural products.

Beta Hatch moved its operations in 2021 from SeaTac, a commercial area south of Seattle, to Cashmere, a small town in the eastern foothills of the Cascade Mountains about halfway between Seattle and Spokane. Emery finds this region to be more affordable for her larger production facility, in part because it receives cheap electricity, three cents per kilowatt-hour, from the federal Bonneville Power Administration. The rural area, she adds, also is "a huge asset in our recruiting. The talent we want to find especially in agriculture is not always used to living in a big city. They are often more accustomed to rural areas. So attracting talent from parts of the Midwest or South into a big giant expensive city is a lot harder compared to a lateral move."[10] Beta Hatch's new location, moreover, appeals to those who enjoy skiing, rock climbing, and other outdoor activities.

The relocation, however, occurred at a difficult time for Emery, who delivered her first child as the team started constructing the new facility and as the COVID-19 pandemic hit. "I've been forced to become more efficient as a leader and to wrestle with a lot of tough issues around sexism and startups," she says. "I think with the pandemic shining a light on work-life balance challenges, more people are recognizing a lot of the continued and systemic challenges for women leaders, so I am excited to blaze a path forward for other female entrepreneurs."[11]

Beta Hatch is one of several insect entrepreneurships that work across the world and grow diverse bugs. Texas-based EVO Conversion Systems favors the larvae of black soldier flies (*Hermetia illucens*); the immature insects eat agricultural waste voraciously and become a source of protein for livestock and aquaculture. Among other bug-farming start-ups are South Africa–based AgriProtein, which grows maggots, producing what the company calls MagMeal; Canada-based Enterra, which is constructing three new production facilities in North America; and Ÿnsect, a French mealworm firm, which raised $125 million in February 2019.[12]

Emery maintains innovation is key to alternative protein fabrication. "There's a misconception on Wall Street and in Silicon Valley that farming is easy and not technology-driven," she notes. "We're pursuing patents on several pieces of equipment specific to producing insects."[13]

As Beta Hatch's output increases, marketing becomes a larger challenge, since this new insect-protein industry enjoys no existing supply chain. "You can't go to a feed producer and buy insect feed," complains Emery. "But if [we] can produce a large insect biomass and sell it at a cost that competes with other ingredients then there's an infinite opportunity."[14]

In Emery's mind, there is no doubt that Beta Hatch will become a billion-dollar company.[15] "Most people don't think about what the fish or chicken they're eating is eating," she explains. "Animal feed is a $400 billion global industry. We're [also] focused on replacing fish meal. That's a $17 billion market."[16] Where she once saw pests, Emery now sees hope; she declares her bugs will "revolutionize the animal feed business."[17]

CHAPTER 6

Leonard Lerer, Back of the Yards Algae Sciences—Growing Algae and Mycelia

As with insects, the thought of consuming algae causes American eaters to squirm, with visions of swallowing pond scum or slimy green goo. Yet these organisms promise to deliver proteins sustainably. Among the planet's original single-cell living organisms, algae remain an abundant food source for hundreds of species. Processors, meanwhile, insert algae-based oils into cosmetic creams, flavorings, colorants, ice cream, and vitamin supplements—in what has become a $5 billion industry. Leonard Lerer, founder and CEO of Back of the Yards Algae Sciences, thinks bigger.

Algae, which Lerer calls "the source of life," enjoy numerous attributes.[1] They grow ten times faster than soybeans, and they need one-tenth of the land area to produce the same amount of plant material. They thrive virtually anywhere, including nonproductive and nonarable land; they produce multiple yields throughout the year; and they do not compete with other crops for soil. They produce no waste material. Pound for pound, they contain twice the protein of meat, more beta-carotene than carrots, more iron than spinach, and substantial amounts of omega-3 fatty acids. Farming algae does not cause erosion or require synthetic fertilizers or pesticides. The organisms do not even demand fresh water, with the protists being quite happy in brackish water, seawater, or wastewater.

Lerer launched Back of the Yards in late 2018, calling it an industrial biotechnology company. Its initial product was an algae-based blue food coloring, but the start-up's mission quickly expanded to "doing protein in a sustainable, cruelty-free, and zero-waste fashion."[2] Motivated to overcome the cultural and environmental destruction associated with industrial agriculture, Lerer, who claims to be a recent convert to sustainability, says of his mission, "Nobody knows until you do it, so let's do it."[3]

Algae help solve several of our greatest population and environmental challenges. Rich in vitamins, minerals, and proteins, they fortify food and combat malnutrition among a growing population. Lerer increasingly mixes algae with mycelia, threadlike filaments from mushrooms and other fungi. By replacing soy protein in cattle and poultry feed, the combinations reduce farming's use of toxic pesticides and curtail the clear-cutting of trees needed to make room for large monoculture plantations.

Lerer claims mycelia offer a unique means to help plant roots receive nutrients, and he calls algae the most efficient means to convert carbon dioxide, sunlight, and water into nutritious food and, in the process, capture greenhouse gases. He suggests algae love "eating" carbon dioxide, creating a self-contained system with built-in carbon capture. He points to Sweden's Algoland project, which on a large scale uses naturally occurring algae from the Baltic Sea to seize a cement factory's carbon dioxide wastes before they're released into the atmosphere;[4] after the protein-rich algae dry, they become feed for chickens and fish.

Algae include a broad collection of organisms, perhaps more than one million varieties, of which only 60,000 have been studied and just 20 used in modern food chains. "There's so much untapped potential," says Lerer.[5] Depending on what he's trying to produce—such as fish food, cultured-meat serum, heme for plant-based meats, or dietary supplements—the entrepreneur uses microscopy and metabolite analysis to select the most appropriate algae and mycelia strains, and he sometimes breeds different eukaryotic organisms to optimize his outputs.

The use of algae as a feedstock has been hyped previously, but it suffered two setbacks during the second half of the twentieth century. In the 1950s, researchers touted *Chlorella pyrenoidosa* as a solution to global malnutrition, with the United Nations suggesting it was "the most ideal food for mankind" and the Food and Agriculture Organization of the United Nations calling it "the best food for tomorrow." Yet no one mastered making *Chlorella* economically on a large scale.[6] During the first oil crisis in the early 1970s, investors tried to convert algae into a biofuel that would replace gasoline, but, again, the process proved to be too expensive.

Lerer, however, argues algal science has progressed substantially and the combination of global warming and population growth creates a new imperative. He believes the biggest potential is actually not in the best-known algae, such as seaweeds and spirulina, the dried blue-green organism that is a common dietary supplement and food additive. More promising is growing algae to be a feedstock for livestock, as well as the bedrock of a sustainable food industry that produces protein-rich alternatives to meat.[7]

Lerer uses his biodigester's mixture to develop a serum in which cultured meats grow. "We want to work towards a future where a leftover piece of food could turn into the meat we would eat tomorrow," one of Lerer's colleagues explains. "This is why we are focused on developing an algae stimulant to help cell-based products grow faster, get more yield, better shelf life, and flavor."[8] Lerer also develops algae-based heme that could serve as the foundation for plant-based meat substitutes like those produced by Impossible Foods or Beyond Meat.

Leonard Lerer, CEO of Back of the Yards Algae Sciences. *Credit: Back of the Yards Algae Sciences.*

Taking a jab at the Western diet's dependence on cattle and pork, Lerer named his firm after the Union Stock Yards once located in southern Chicago and featured in Upton Sinclair's *The Jungle.* Back of the Yards Algae Sciences, in fact, is one of several food-based start-ups located within a 100,000-square-foot former meat-packing facility. "The meat of hundreds of millions of pigs was probably processed in here," explains Lerer. "And soon we will be able to make a burger that is 100 percent algae-based and cruelty-free."[9] According to the CEO, the building has gone full circle as a center for food innovation, starting early in the twentieth century when it launched canning, cold storage, and industrialized meat production. Today, while it maintains the smoke-stained

bricks from its former pork production, the structure serves as a laboratory for alternative proteins, while large-scale production of algae- and
mycelia-based colorants, feed, and food is done at a facility in Utah.

Lerer fills his laboratory with state-of-the-art scientific equipment, as
well as a gaggle of bottles and pipes containing green goo in different
stages of liquidity. One process begins with adding fast-growing algae
to decomposing waste materials in an anaerobic (without oxygen) biodigester. After seven days, Back of the Yards "harvests" the algae using a
patent-pending process that removes the lipids, or fats, and then extracts
what Lerer needs to produce feed for fish and animals, fertilizers for
plants, or oil for frying. His procedure, fortunately, eliminates algae's
typically unpleasant smell and taste.

When provided to cattle, the algae-based protein feed increases the
animals' weight.[10] When applied at fish farms, it replaces conventional
fish meal—typically a combination of anchovies, herring, menhaden,
and other small, open-ocean fish—whose mass harvesting destroys
marine ecosystems. According to the consulting firm Oliver Wyman,
"the cost of farming algae in most locations is between $400 and $600
per metric ton, a 60 percent to 70 percent savings compared to fishmeal,
which costs $1,700 per ton."[11] Algae, moreover, grow reliably, reducing
a fish grower's risk of not catching enough small fish to supply her carp,
tilapia, salmon, and catfish farms.

Lerer's latest targets are vertical farms, those indoor plant factories
profiled in chapter 7 that optimize environmental conditions, nutrition,
and lighting. Such sophisticated greenhouses allow vegetables and fruits
to be grown locally, reducing the need for transcontinental shipping
and limiting the use of chemical pesticides. After two years of research,
Lerer discovered how to create an extract from blue-green algae—specifically the protein-pigment complex known as phycocyanin that regulates photosynthesis in the algae—to stimulate the produce's growth,
yield, and quality. His initial tests found that the biostimulant reduced

maturation times by 21 percent, produced 12.5 percent more lettuce, and postponed wilting by several days.[12] According to Lerer, the algae extract "gives vertical farmers a new avenue for reducing costs, while improving produce's color, vigor, nutrient content, and preservation."[13]

Lerer's business plan is to sell to companies that market to customers rather than try to approach consumers directly. As examples, he offers spirulina-based flour to bread and pasta producers; supplies algae-based ingredients to yogurt, cheese, and chocolate makers; and provides to vertical farmers an algae extract that stimulates their plants' growth and yield. Lerer also wants to develop "hybrid foods," good-tasting meals that we've never imagined, such as cell-cultured shrimp spread on plant-based pasta extruded from 3D printers.

Lerer, like all entrepreneurs, struggles to attract investments and scale up production. He claims his opportunities to be massive, but he knows he must advance the science, obtaining, for example, the right ratio of fatty acids, such as omega-3s, to replicate the natural balance found in fish meal and fish oil. He views himself as more a researcher than a manager; despite establishment credentials—he holds joint MD and MBA degrees, taught college-level pathology and epidemiology, and advised deep-pocketed financiers about life science opportunities—he refers to himself as an outsider and disruptor.

Lerer loves giving tours of his equipment, exudes confidence in his experiments, and argues that Back of the Yards Algae Sciences is part of the revolution rattling industrial agriculture. "What could be more transformative?" he asks. "We're producing a protein-rich food as a healthy alternative to meat as well as tackling climate change by gobbling up carbon dioxide."[14]

PART 2

REDUCE FOOD WASTE

Of all the issues discussed in this book, food waste seems at first glance the least ripe for a technological fix. After all, a big part of the problem is behavioral: people buy too much food and throw too much away. Yet habits are often harder to change than technology. Governments for years have struggled to reduce food remains, and more recently, reformers have turned to financial incentives, public information campaigns, farm-to-table markets, and online shopping as solutions to the mountains of discards. These efforts certainly play a role, but few expect chefs to serve smaller portions or eaters to devour everything on their plates and in their refrigerators. Instead, entrepreneurs see profits in devising ways to cut waste before it gets to consumers' trash bins. It is these measures that may ultimately make a dent in the alarming statistics.

Those numbers are particularly dire in the United States, where Americans waste nearly 40 percent of our food, making us the global leaders in thrown-away produce, cereals, and meats.[1] Discards annually cost $218 billion to produce and weigh 80 billion pounds, about equal in heft to one thousand Empire State Buildings.[2] Our wasted food, about 400 pounds per person, accounts for as much carbon pollution as 37 million cars,[3] and its growing requires 780 million pounds of pesticides and 4.2 trillion gallons of water.[4] Farming, processing, transporting, and discarding food we never eat costs a typical American four-person family $1,800 each year.[5] More than 80 percent of what we discard is still wholesome food, but it ends up as the single largest component of dumps in the United States, representing squandered labor, energy, and resources.

Worldwide, food waste costs roughly $2.6 trillion (yes, with a "t"), most of it resulting from inefficient harvesting, poor distribution networks, pests, diseases, and the rotting of perishables on the shelves or in the refrigerators of retailers, restaurants, and consumers. If food waste were a country, it would be the planet's third-largest emitter of greenhouse gases, behind only China and the United States.[6] Of the eighty most effective ways to combat climate change, Project Drawdown ranks reducing food waste number three.[7]

Where food wastes settle also impacts the environment. If composted, they become a dark mix that enriches soils, but most discarded food gets dumped into landfills, where organic materials decompose and vent methane, a powerful greenhouse gas.

Individual consumers are not all to blame. Part of the food-waste problem results from our extended supply chains. Much of the United States' lettuce and other greens grow in the sunny fields of California and Arizona and travel long distances, taking weeks to get to markets in the Midwest and East. Despite refrigerated trucks and trains, some of that food rots during the crossing, and its dietary value falls substantially. Produce, in fact, loses 30 percent of its nutrients in a short three days after harvest, and up to 55 percent of its vitamin C disappears within a week.[8]

The COVID-19 pandemic highlighted cracks in our food system that allow food waste to increase. According to S2G Ventures, it "showcased some of the tradeoffs we have made over the last century—efficiency at the cost of resiliency; scale at the expense of variety; price at the expense of value; globalization at the expense of our local communities."[9]

No doubt there are valiant efforts around the world to make the system more resilient and less wasteful. By charging a fee for spoilage and placing automated disposal bins near apartment buildings, South Korea increased its composting rate from 2 percent in 1995 to 95 percent in 2020. New York runs the United States' largest composting program, collecting food scraps from residences and restaurants, but it still sends more than 1.3 million tons of food waste to landfills every year.[10] A few cities encourage supermarkets and chefs to donate still-edible food to charities, yet safety standards and complex supply networks complicate such noble work.

To reduce the waste associated with long-distance transport, many small-scale growers offer vegetables, fruits, and other foods to farmers' markets and farm-to-table restaurants. An example of a localized food and aquaculture effort is Superior Fresh, a ten-acre family farm in Wisconsin's Coulee Region that annually grows 1.8 million pounds of leafy greens and 160,000 pounds of Atlantic salmon and steelhead trout. At a warehouse near the L train stop at Brooklyn's Montrose Avenue— not your typical farm country—Edenworks raises salmon, shrimp, and striped bass on its lower level while growing baby greens and micro-greens upstairs. The US Department of Agriculture calculates that the number of farmers' markets reached 8,771 in 2019, up by 43 percent since 2010,[11] yet many such outlets are closing or cutting back because, according to National Public Radio, "there are too few farmers to populate market stalls and too few customers filling their canvas bags with fresh produce at each market."[12]

Online grocery shopping has been proposed as another means to reduce food waste, although some argue meal deliveries lead to more throwaways. With digitalization advances and COVID-19 pressures, projections estimate such sales will grow to $143 billion by 2025, which represents 30 percent of omnichannel food and beverage spending.[13] Given that earlier high-profile and well-funded food-delivery efforts, including Peapod and Webvan, scaled back or closed their doors after only a few years, these numbers suggest the food retail industry is changing at an unprecedented rate.

In short, current methods of reducing waste go only so far, and much more is needed as consumers adjust their grocery-buying habits. The problem is crying out for new solutions and technologists are eager to heed that call, with a little financial benefit for themselves. Also driving their innovation, says farm analyst Seana Day, is "the desire for better tracking, transparency and security in our food system."[14]

In the chapters that follow, we will meet entrepreneurs who are working to address these issues. An Amazon veteran builds vertical farms that grow vegetables and fruits in urban warehouses close to consumers. Biochemists produce an acidic solution to block the enzyme, known as polyphenol oxidase, that causes fruits and vegetables when exposed to air to change color and get tossed out. An Intel alum builds autonomous robots that pick strawberries more efficiently. A former banker deploys blockchain technology to track a crop's every movement, ensuring transparency and traceability. One entrepreneur produces biodegradable plastic packaging, while another 3D prints meals sized for each individual's appetite. Their innovations promise to cut our staggering waste, allowing us to provide good food to more people with less pollution.

CHAPTER 7

Irving Fain, Bowery Farming— Bringing Crops Closer to Consumers

Irving Fain from an early age displayed a knack for commerce. In elementary school, he bought and repackaged toys and then sold them to his classmates at a markup. "I've been an entrepreneur since I was a young kid," he says, "and I chased every little hustle growing up."[1] After graduating from Brown University, he built iHeartRadio for Clear Channel Communications and then cofounded a marketing analytics start-up that was sold to Oracle.

With money in his pocket and wanting to do something on his own, Fain says, "I spent a lot of time looking at various opportunities, and the more time I spent learning about and researching the agricultural system, the more I realized that agriculture was at the epicenter of so many of our global issues."[2] Having grown up in Providence, Rhode Island, and realizing that 70–80 percent of the world's population soon would live in cities, Fain decided to focus on urban farming. Also motivated by his mother, an avid chef with a passion for quality and locally grown food, he concentrated on providing fresh and nutritious produce to city residents.

Fain cofounded Bowery Farming in 2014, proclaiming it a tech company that is "thinking about the future of food."[3] He refurbished an abandoned warehouse in Kearny, New Jersey, an industrial area once known for building warships, to become a vertical farm that grows vegetables and fruits indoors. He claims to be rethinking farming, growing plants closer to consumers, democratizing access to quality food, and producing crops in layers rather than rows.

The facility looks nothing like a traditional farm or even a standard greenhouse or hoop house. To enter the large building, you must don a disposable sterile gown, slippers, and hairnet, and then cross through a pressurized air lock that blocks pests and therefore eliminates the need for expensive and dangerous pesticides. There is no soil in this farm, since the crops sprout in nutrient-rich waterbeds piled from floor to ceiling; misters spray specialized nutrients onto the roots. Sensors and cameras seem ubiquitous, tracking how each plant responds to airflow, carbon dioxide, and light. Proprietary software individualizes conditions throughout the warehouse, so a certain batch of kale gets more water or a bunch of lettuce obtains warmer temperatures. Photosynthesis-prompting light comes from light-emitting diode (LED) lamps rather than the sun; using machine learning, those lights provide the precise wavelength needed by each plant.

Farmers have tried greenhouse-based hydroponic agriculture since the 1930s, producing numerous commercial crops, including strawberries, tomatoes, and herbs. Yet unlike Bowery's technology-focused vertical farm, those greenhouses relied on sunlight, planted across a horizontal plane, and lacked robotics and artificial intelligence.

Fain automates all of Bowery's operating systems—those that plant, nourish, water, and harvest—and an integrated platform of sensors, control systems, cameras, and robots is designed to optimize the farming processes. According to the entrepreneur, "We deeply believe in the power of technology to make drastic, necessary improvements to the food system."[4]

"Building technology is hard, it's expensive, and it takes time," says

Bowery Farming warehouse. *Credit: Bowery Farming.*

Fain, "but the tech you use in indoor ag has a direct and clear impact on the economics of the business you're creating, the varieties you can grow, and the efficiencies you can generate, and we realized this early on."[5]

One of Fain's initial hires was a mechanical engineering graduate of Penn State University with an MBA from the MIT Sloan School of Management and seven years of experience at Amazon helping to design automated fulfillment centers where robots pick and pack a seemingly endless stream of orders. Brian Donato, who has since left Bowery, claims the vertical farm—with its multiple layers of modular trays—looks and operates like a giant computerized fulfillment center, but one growing lettuce, arugula, and bok choy rather than stacking books, clothes, and electronics.

Donato brought to Bowery an appreciation for the power of robots. When he began at Amazon, humans staffed the fulfillment centers, but the e-commerce company in 2012 purchased Kiva Systems, which makes machines that automate many warehouse operations. Soon after integrating Kiva's logistics, Donato moved to Amazon Fresh, the firm's online grocery, where he added more intelligence and speed to the computerized delivery system, since "everything needed to be inspected and delivered within certain time windows, when produce is at its best."[6]

Irving Fain, CEO of Bowery Farming. *Credit: Bowery Farming.*

Fain sees those kinds of efficiencies as key benefits of vertical farming. Bowery, he boasts, produces one hundred times the output of an equivalent-size plot of horizontal land. The hydroponic system delivers targeted nutrients to diverse plants, curtailing the need for synthetic fertilizers and pesticides, and it relies mostly on recycled water, using a small fraction of H_2O compared with traditional agriculture. Since these systems don't rely on soil, they also may be able to claim organic status earlier than conventional farms that must wait three years for synthetics to wash away from their soil, perhaps increasing the availability and lowering the costs of organic produce. Also significant, Bowery's growing season never ends, allowing the delivery of fresh and organic greens to local restaurants and supermarkets—such as Whole Foods Market, Ahold, Amazon Fresh, and Westside Market—even during February snowstorms. By avoiding hurricanes and droughts, made more pronounced by climate change, climate-controlled vertical farms are the ultimate defense against global warming, Fain continues, and they can operate in Alaska, the Sahara Desert, and other hostile environments.

Beyond the environmental advantages, Fain sees economic and social ones. He asserts proudly that high-tech farming provides well-paid jobs in formerly abandoned spaces within neglected neighborhoods, and the local labor force needs no agricultural experience.

Vertical farming, moreover, regulates and optimizes all growing conditions, ensuring the produce consistently tastes and looks great. With filtered air and pressurized buildings, these warehouses, compared with conventional farms, offer a safer and more secure place to work and grow crops. Vertical farms, says Fain, also provide more food varieties, reversing the decades-long farming trend toward standardization, and they offer a controlled space in which to grow more nutritious greens made possible by advanced breeding techniques; produce, says Fain, is becoming the modern pharmacy in which food and health care converge. Indoor agriculture also better adapts to economic shocks and changing market conditions; during the COVID-19 pandemic in 2020, for instance, Bowery increased its growth by 600 percent in response to large jumps in online orders and retail sales.

Bowery in 2017 launched its first farm in Kearny, which it later converted into a research and development center, and the next year it built a commercial facility in the same area but thirty times larger. Its unit in Baltimore, constructed on a former farm and completed in 2019, offers another 3.5-fold increase and operates on clean hydropower. In late 2020, the company began constructing an even larger and more technologically advanced vertical farm in Bethlehem, Pennsylvania. While the initial New Jersey operation features one warehouse-size room, the newer facilities offer multiple growing chambers, where cilantro enjoys hot and dry temperatures while bok choy thrives in cooler and moister climes. As his vertical farms proliferated, Fain opened a corporate office south of Midtown Manhattan.

No doubt vertical farming faces challenges, including the high up-front costs associated with racks and sensors as well as the ongoing expenses associated with sophisticated lighting and ventilation. Compared with midwestern spreads, land near urban centers tends to be expensive, and labor costs are high. Although lights and air filtration systems run almost constantly, Fain maintains that vertical farms provide benefits to the electric grid and the planet; since the crops do not care when "night" begins, Bowery responds to electric-utility price signals and reduces power demand during times of peak prices and high pollution. The firm also plans to expand its use of renewable resources and cogeneration to power all its warehouses.

Vertical farms may not be practicable for every crop, at least currently. Most ideal are high-value produce, such as salad greens and tomatoes, which do not store well and must get to markets quickly. Less economical are commodity crops, such as corn and rice, which can be stockpiled and shipped over long periods.

A few vertical farms initially confronted financing challenges. PodPonics could not raise enough capital to expand, so it closed its doors, and New Jersey–based AeroFarms predicted in 2015 that it would build twenty-five indoor farms by 2020 but completed only two. Yet Bowery's investors include Google; First Round Capital; chef Tom Colicchio; Temasek, Singapore's state fund; and Henry Kravis, the founder of private equity giant KKR. Plenty, a San Francisco–based start-up, raised a substantial $226 million from venture capitalists and constructed large vertical farms in San Francisco and Los Angeles.

Indoor farms—sometimes referred to as "controlled environment agriculture"—attracted more than $1 billion of investment in 2020,/ and they grow everything from leafy greens to cannabis, flowers, herbs, and mushrooms. In Cincinnati, 80 Acres Farms built a fully automated facility about the size of two and one-half football fields. AppHarvest operates two sixty-acre greenhouses in Kentucky, within a half day's drive of markets in the Northeast and Midwest. Florida-based Kalera erected a vertical unit near Orlando International Airport that annually produces six million heads of lettuce, and it plans new facilities in Columbus, Ohio; Atlanta; Houston; Denver; and Seattle. IKEA sells, at many of its store restaurants, lettuce grown on-site, often within truck-size containers. Growing Underground, a start-up based in Clapham, in South London, cultivates herbs and vegetables hydroponically within a maze of tunnels that were World War II bomb shelters. Electronics giants Panasonic and Sharp, better known for manufacturing television screens, launched indoor vegetable farms in Singapore and Dubai using specialized LED lamps to optimize photosynthesis.

Lowering costs is key to vertical farming's future. Robots dominate the work, but salaried and hourly workers still direct the machines, maintain quality control, package the fruits and vegetables, and clean the equipment. Although the recent 85 percent reduction in costs, and a doubling of efficiency, in LED lighting made vertical farming possible, food warehouses, if they are to be economical and sustainable, must further increase those efficiencies as well as switch to low-cost, renewably sourced electricity. According to Fain, Bowery's lighting expenses must be less than the fuel costs associated with transporting produce across the country conventionally. The CEO also anticipates continued advances in Internet of Things (IoT) capabilities and breeding platforms will further increase the attractiveness of indoor farming.

Fain acknowledges that vertical cultivation must evolve further. "Our sector is in its infancy," he says, "and we too have work to do in curbing our own emissions, investing in alternative energy sources, moving away from plastic packaging, and selling beyond just leafy greens. Fortunately, these are all areas that we're aggressively making progress toward at Bowery. We have seen meaningful gains on many of these fronts already, and I am not dissuaded by the work ahead; in fact, I'm energized by the opportunities that are still in front of us."[8]

A few academics and skeptics disagree, question the economics of vertical farming, and claim it will remain a niche means to grow a few high-cost vegetables. Yet even large-scale, industrialized growers admit that climate change prompts them to look indoors for at least a portion of their plantings.[9]

The hard-charging Fain—whose morning workout includes running, surfing, swimming, or boxing—claims vertical farming is transforming agriculture. He argues that Bowery's continued improvements will provide affordable and nutritious food—grown locally and sustainably—to expanding urban populations. His own personal story, moreover, shows how agricultural innovation attracts hustling, city-born entrepreneurs who appreciate the power of technology.

CHAPTER 8

James Rogers, Jenny Du, and Louis Perez, Apeel Sciences—Coating Foods

James Rogers, Jenny Du, and Louis Perez want to cut food waste in half. With PhDs in materials science and chemistry, they devise plant-based coatings that double the shelf life of avocados, oranges, and other produce. "If we wasted less food," they say, "we could feed more people. But how to stop food from going bad?"[1]

The problem of food loss, the scientists claim, is rather straightforward: "The two leading causes of produce spoilage are water loss and oxidation—that's water evaporating out of the produce and oxygen getting in."[2] So the trio creates tasteless, odorless, invisible, and edible coatings—consisting of fatty acids and other organic compounds extracted from the peels and pulp of produce—that act as physical barriers to keep water in and oxygen out. To improve and market that innovation, in 2012 they formed Apeel Sciences, headquartered in a one-story warehouse in a new industrial park in Goleta, a vibrant community of technology upstarts near the University of California, Santa Barbara (UCSB).

Rogers initially thought he would use his doctorate to create a paint that converts the sun's rays into electricity, providing less expensive power than photovoltaic panels; remembering his long hours testing alternative stains, he describes that period of his life as "watching paint dry." One day while driving between the University of California's labs in Santa Barbara and Berkeley, he marveled at the expansiveness of the Salinas Valley's rich farmland but, while listening to a podcast about world hunger, wondered why so many people go hungry amid such food abundance. "The problem," he eventually concluded, "came down to distribution. We couldn't get the food that was being grown to where the people were who need to eat it. And so I was curious what precludes us from distributing food and it all comes down to the notion of perishability."[3] He decided to stall spoilage.

As a materials scientist, Rogers knew coatings protect steel from rusting, so he felt a similar barrier should retard the rate of food rotting. After jotting out a plan on a napkin, he phoned his mother and said, "Hey, Mom, I got this idea for a company," to which she responded: "That sounds really nice, but you don't know anything about fruits and vegetables."

Rogers admits Mom was right, so he began to read up on plant biology and wrote a proposal to the Bill & Melinda Gates Foundation to reduce postharvest food loss in developing countries that lack access to refrigeration. Its $100,000 grant in 2012 allowed him to join with two other researchers—Jenny Du and Louis Perez—and start Apeel Sciences.

Perez obtained his doctorate in materials from UCSB and focuses on research and development, engineering, and software, having gained six patents for technology he developed at Apeel. Du grew up in Canada in what she calls a "super blue-collar family."[4] Her parents fled Vietnam by boat in the late 1970s, and even though her father reached only the sixth grade, they pushed her to study engineering chemistry at Queen's University in Ontario, Canada, and then obtain a PhD in chemistry from UCSB. None of Apeel's founders, she observes, "were in fresh fruit or vegetables or food or plants to begin with, at all."[5]

Jenny Du, James Rogers, and Louis Perez (left to right), cofounders of Apeel Sciences.
Credit: Apeel Sciences.

Perez, Du, and Rogers say the harder transition was from academics to entrepreneurs, from advancing a concept to producing a product. Trained as scientists, they admit, they're "pretty married to the beauty of the idea, the elegance of the science and the idea, but then, to actually make it work in a business setting, that's the first one percent. To actually carry over from . . . an idea into true innovation and application, requires a whole bunch of other stuff." Rather than be locked into your original plan, they say, you have to "listen to the feedback from your customers, from your stakeholders, as to what might alternatively be possible or help it gain traction."[6]

To attack the problem of spoilage, Rogers, Du, and Perez advanced a straightforward-sounding idea: to produce coatings—what they call "protective peels"—that allow a variety of produce to stay fresh two to three times as long.[7] Their challenges, however, were anything but simple—to invent the coatings, design an application process, make the technology affordable, and convince customers to use it.

Apeel's founders take advantage of cutin, a natural sealant plants use to retain moisture, but they create proprietary mixtures of lipids and glycerolipids from the peels, seeds, and pulp of various fruits and vegetables that, when water starts evaporating, join together to build a film that traps moisture and rejects oxygen. "Part of [the mixture] really likes water, and part of it really doesn't like water, which means you can get some limited solubility of that material in water," the trio says. "Once they dry, then they have the ability to block water."[8]

Since different types of produce degrade differently, the innovators began investigating why a strawberry naturally lasts a couple of days while a lemon retains its fruitiness for several weeks. "Remarkably," they say, "it's not that they're made of different things, but rather that the molecules on the surface of a strawberry are arranged dramatically differently than they're arranged on a lemon." Apeel's challenge was to tailor its formulas for each type of produce in ways that augment the natural protective barriers of the vegetables and fruits. Although different plants offer richer sources of lipids, coatings can be made from any peels and seeds that remain unused after harvesting. "Essentially," the founders say, "Apeel is cutting and copying from what the natural world is already doing,"[9] and the start-up builds "a world that works with nature, not against nature."[10] Because "we use food to preserve food," they conclude, both traditional and organic farmers can utilize Apeel's products.[11]

The young entrepreneurs recognize that humans have tried for centuries to protect fruits and vegetables, without much success. Medieval monks dipped apples in beeswax, a shield that made fruit look good for a short period but did little to retard dehydration and oxidation. The difference is that we now understand the detailed science of how produce ripens and rots.

"Produce is a living, breathing thing, even after it's picked," explain the founders. "The trick is to maintain as much moisture as possible and maintain a very delicate balance between the rate of oxygen getting in and the rate of carbon dioxide getting out. If you cut down the oxygen too much, the fruit won't develop appropriately and it will develop off flavors, and if you don't cut down the oxygen enough, you'll have no impact."[12]

Apeel markets its product, called Edipeel, as a powder that its partners—fresh-food suppliers and retailers—mix with water and then spray, dip, or brush onto produce. "The result," says Rogers, "is that we form this special structure, this special barrier, which mimics that structure which is employed by longer shelf life produce."[13] With oxygen repelled, the fruits and vegetables become less stressed, their metabolic rates slow, and they better defend themselves against biotic infections such as mold.

Apeel started with avocados, which ripen and brown quickly and often end up in trash bins. When coated and given a longer shelf life, the often-imported fruit can be transported by sea without refrigeration, rather than by air, thereby reducing greatly the product's cost and carbon footprint. Rogers, Du, and Perez subsequently developed protective layers for limes, mandarins, oranges, and organic apples, but they have set their sights on all fruits and vegetables.

Initial customer reviews judge the edible "skin" to be tasteless and colorless, neither waxy nor greasy. Managers at Costco, Kroger, and other grocery stores say the coatings allow them to sell more fruits and vegetables, face lower waste-management costs, and suffer fewer customer returns. A former CEO of Whole Foods Market commented, "As a fresh grocer for over 40 years, I can truly say that I have never seen anything like Apeel, and I believe it has the potential for global impact on the food system."[14]

Consumers also appreciate the reductions in packaging waste. Responding to their pressure, Walmart in September 2020 ditched its plastic shrink-wrap around cucumbers; the giant food marketer estimates that coating rather than wrapping 100,000 cases of the vegetable removes the equivalent of 165,000 single-use plastic water bottles from landfills.[15]

Initial reactions from investors also are impressive. Apeel raised $250 million in capital in May 2020, bringing its total valuation above $1 billion and making it another food-focused unicorn. Building on its early grant from the Gates Foundation, Apeel in late 2020 raised another $30 million to help smallholder farmers in remote locations, such as sub-Saharan Africa, obtain access to new markets and retail opportunities.[16] Oprah Winfrey, one of the firm's high-profile investors, comments, "I hate to see food wasted when there are so many people in the world who are going without. Apeel can extend the shelf-life of fresh produce, which is critical to our food supply and our planet too."[17]

Edipeel does not eliminate food waste. Rough handling or poor refrigeration can pierce the coating and expose fruits and vegetables to contaminants. Restaurant patrons also may not eat their helpings, and forgotten produce at the backs of refrigerators can decay over time. Yet Apeel claims to be "the only postharvest solution that creates an optimal microclimate inside every piece of produce, which leads to extended shelf life and transportability."[18]

Consumers enjoy fresher, better-tasting, and more nutritious produce. Distributors spend less on packaging and controlled-atmosphere storage. Grocery stores reduce their discards and thereby cut their costs.

Landfills with less organic wastes emit less greenhouse gases. Finally, by slashing food waste, the overall agricultural system expends less on water, wages, plastics, and chemical additives, as well as ensuring more food gets to more people.

Apeel's cofounders expect that in twenty years, sophisticated coatings will almost eliminate food waste and increase food choices. "You'll be able to walk into any corner store, any little bodega, any 7-Eleven, any mom-and-pop store," they predict, "and be able to pick up a piece of produce that was grown by a small farmer on the other side of the world, and it will be better than any produce you've ever eaten in your life."[19]

CHAPTER 9

Bob Pitzer, Harvest CROO— Picking Strawberries Robotically

Bob Pitzer, another newbie to agriculture, teamed up with Gary Wish-natzki, a third-generation grower at Wish Farms, appropriately located in Plant City, Florida, to harvest strawberries more efficiently. Before bringing automation to farming, Pitzer helped produce *BattleBots*, a television show featuring warrior robots, as well as assist Intel mechanize its fabrication of microchips.

In 2013, the pair formed Harvest CROO Robotics, in which the acronym stands for "computerized robotic optimized obtainer." Wish-natzki understands that berry picking must modernize because it requires backbreaking work done by a dwindling workforce, and Pitzer hopes to utilize his experience to automate farming profitably.

Pitzer says that before Harvest CROO Robotics, no one had devised a selective picker for fruits or vegetables that are vulnerable to squishing and bruising. The task is complex. "You go out and look at a straw-berry plant," he explains. "It might have ten berries underneath it, but you might only want [the three ripe ones]. You have to be able to decide which three. There's an intelligence factor built into that."[1] Pitzer argues that autonomous strawberry picking requires something far more ingenious than typical robots, which repeat the same actions with no variation, whether they are welding a car frame, assembling a smartphone, or making microchips. Agriculture, in contrast, imposes a multitude of variables, including changing light and humidity, wind and rough terrain, different shapes and shades of leaves, and berries with evolving ripeness.

While Harvest CROO focuses on making robots for traditional fields, Pitzer acknowledges such machines are particularly suited for indoor farms, such as Bowery Farming's warehouses, described in chapter 7, that feature more controlled and systematized environments. Although Pitzer initially focused on strawberries, because the fruit must be harvested regularly throughout a season, he expects his "production-level techniques and supply chains can be swiftly converted to other crops."[2]

The robotics engineer devised numerous versions of a GPS-guided and self-driven machine that views and evaluates individual plants and then delicately plucks selected berries from their vines. Taking advantage of advances in visioning and computer processing, he deploys multispectral cameras to snap two hundred images each second, building profiles of every plant in the six-hundred-acre farm.

Pitzer's B7 iteration weighs about 25,000 pounds, stretches thirty feet, and looks like a truck trailer; he explains that the harvester must be big because "you are going to have thousands of pounds of berries. And you need a large machine to physically carry those berries out in the field."[3] The apparatus features sixteen robotic "plucking" arms, cameras shielded from the fluctuating light of sun and clouds, and lasers that sense obstacles, allowing the behemoth to operate twenty-four hours per day and seven days per week. Evening picking offers several advantages: since cooler berries become a bit harder, or less squishy, they are easier to pick and less likely to bruise; chilled fruits, moreover, require less energy to cool and store. Operating constantly, the machine increases a farm's output and ensures timely harvesting; according to Wishnatzki, "Right

Bob Pitzer, cofounder of Harvest CROO. *Credit: Joel Meine.*

now our workers don't always show up on Saturday and Sunday," and because of labor shortages, some fields are never picked.[4]

The robot, nicknamed Harv, spends eight seconds over twelve rows of strawberries, about the same amount of time that a human farmworker takes to deal with a single plant. It scans with stereoscopic cameras equipped with multispectrum and infrared vision and draws a 3D map of each vine. The system's algorithms take the enormous amount of data and target berries with the best color and size. Pitzer boasts that the robot's accuracy in picking ripe fruit reaches 98 percent, thereby reducing waste.

"Stereovision has been around for decades," says Wishnatzki, "but the difference now is the processing power that has made all of this possible.

Harv, Bob Pitzer's B7 harvester. *Credit: Harvest CROO.*

Just in the last five years, we're processing something like 30 gigabytes a second and doing hundreds of image pairs. It's an amazing amount of computing power. Twenty years ago that just wasn't possible."[5]

Harv's collected images also identify pests and forecast yields. "At Wish Farms, we used to have three or four full-time people that all they do is go out and count blooms and try to determine what our production is going to be like in three or four weeks," says Pitzer. "It had been such a small sample size and inexact science, but we're getting much better forecasting now that we're taking all these images of the plants."[6]

After the robot targets a ripe berry, a food-grade silicone rubber claw cups the selected fruit and pivots to pop it from its runner. The fast-moving arms are part of a patented Pitzer Wheel, which rotates rapidly to pick and move berries. The wheel delicately transfers a plucked fruit from one of the soft claws to a belt, on which more cameras inspect the fruit a second time, and another robotic arm then places quality strawberries into consumer-ready packs.

The Pitzer Wheel "really gets us to the commercial speed we need in picking," says Wishnatzki, who calculates that early versions increased yields by at least 10 percent. "It's not a multi-access robotic arm that has to grab something and move it somewhere. The wheel maneuvers the claw to pick the strawberry. The wheel spins to give us the speed to pick the plant much more rapidly."[7] Bob Pitzer, says Gary Wishnatzki, "is changing the future of agriculture."[8]

This machine with a gentle touch then moves on to the next series of plants, hissing as it goes. It selectively picks about five strawberries every second and covers about one-third of the Wish Farm each day. The robot returns to the fields regularly, ensuring that a steady stream of ripe berries makes it to market. In contrast, a team of hired farmworkers would traverse the field over a few days, leaving behind rotting berries that become ripe before or after their efforts.

The owner of six US patents, including for the mapping of individual plants, Pitzer is one of several entrepreneurs developing specialty-crop robots. Agrobot, a Spanish firm, and Dogtooth, in the United Kingdom, test other strawberry pickers, while farmers in the Pacific Northwest deploy an apple-plucking robot with a dozen mechanical arms. Researchers at the University of Florida devised machines that inspect thousands of strawberry seedlings, using artificial intelligence to identify those that will be easiest for a robot to harvest. According to the Bank of America, the agricultural robot market rose from $817 million in 2013 to $16.3 billion in 2020.[9]

Labor issues initially motivated Wishnatzki. US farms face a shortage of seasonal workers, in part because of government crackdowns on immigration and restricted visas for temporary agricultural laborers. About half the nation's 2.5 million farmworkers are undocumented and fear deportation, yet farm owners recognize that their hands keep the US food system working.

Another part of the labor-shortage problem is simple demographics—fewer and fewer people, particularly the young, want to perform the toil, often done stooped over in the hot sun. "The labor force keeps shrinking," complains Wishnatzki. "If we don't solve this with automation, fresh fruits and veggies won't be affordable or even available to the average person."[10] Strawberry picking, moreover, has been called "among the most abusive industries in terms of their treatment of labor."[11] The United Farm Workers of America, whose twenty thousand union members across the country seek better pay and working conditions, argues that machines cannot perform the delicate picking done by skilled humans, and Wishnatzki asserts that a mechanical harvester completes the work of thirty pickers and diminishes the need for field labor; still, the farmer-innovator maintains that a robot offers new and better-paying job opportunities for workers who can program, maintain, and repair the sophisticated machines. Trying to address that need, Pitzer works with Dean Kamen, inventor of the Segway Human Transporter, to teach science, technology, engineering, and math (STEM) to the children of farmworkers and to organize large-scale robot competitions. They also work with FIRST, a nonprofit group that engages young people in mentor-based programs that inspire technology innovation.

Wishnatzki argues that robots bring needed modernization to agriculture. "Over the last 15 years we've heard from growers who are dealing with labor issues and government and saying it just isn't fun anymore," he says. "Our goal is to put the fun back in farming."[12]

Harvest CROO's funding seems to be a group effort within the strawberry industry. Investors include other Florida growers as well as giant processors such as Driscoll Foods, California Giant Berry Farms, and Naturipe—all hoping machines will solve their labor issues, reduce waste, and increase yields.

Robotic pickers, despite their sophistication, face challenges. Cost is most prominent, particularly since farmers with low profit margins resist investing in new equipment. Harvest CROO tries to get around this barrier by leasing its machines at a cost equal to or less than what growers pay for field labor. Wishnatzki expects expenses to fall as technology advances and manufacturing improves.

Pitzer claims to relish the trials, arguing that revolutionizing agriculture is far more interesting than devising television-based robotic warriors. He boasts that Harvest CROO Robotics increases the efficiency of harvesting, enhances agricultural production, cuts food spoilage, and brings nutrients to more consumers. Yet the engineer-entrepreneur dreams larger than a mechanical strawberry picker, envisioning autonomous robots that plant, nurture, and gather all the world's crops.

CHAPTER 10

Raja Ramachandran, ripe.io— Tracking Food with Blockchain

Blockchain has become a poster child of disruptive technology. Bill Gates calls it "a technological tour de force," and one business analyst says it "will do to banks what email did to the postal industry."[1] It promises to challenge aging monopolies and open up new business opportunities.

Raja Ramachandran wants to bring such disruption to agriculture. He believes that digitally tracing all activity from farms to forks can make supply chains more transparent and trustworthy, helping consumers to "know their food." Tracking a crop's journey has the potential to increase efficiency, reduce food waste, and monitor its safety, sustainability, and quality. "By engaging actors throughout agriculture's production and distribution," Ramachandran says, "blockchain empowers them to make confident choices about the foods they grow, sell, and eat."[2]

Put simply, blockchain is a sophisticated ledger. First adopted as Bitcoin in the financial services industry, it represents a series of connected blocks, each of which contains information about a specific transaction. When strung together, those blocks form a chain, a distributed database that holds a multitude of records. Yet unlike a conventional database, in which only a few institutions (such as a bank) control and reconcile information, blockchain offers a shared, open-source digital ledger. To protect the data's integrity, every participant in a transaction can validate each block, which sophisticated cryptography protects. This chronological chain of trades thus provides secure, up-to-date information to all engaged parties.

Ramachandran knows a thing or two about blockchain. As part of a twenty-year career with Citi and other banks, he directed product development at R3CEV, a consortium of such corporations that built and deployed a distributed ledger to streamline financial transactions. As a banker reviewing investment opportunities, he also evaluated how blockchain impacts the health-care, insurance, legal, entertainment, and electronics industries.

Like most of the profiled entrepreneurs, Ramachandran admits to initially knowing little about agriculture. He emigrated from India to the United States as a child, so he had limited interactions with the side of his family that farmed and milled rice. Yet he grew up with his mother's South Asian cooking, and he came to appreciate quality ingredients. After helping to launch R3CEV, Ramachandran says, he "got to the point in my career that I was looking for something more meaningful," so he evaluated opportunities in the various fields his bank had investigated. "Yet perhaps because of my mother's cooking," he says, "I ultimately stumbled into food." Having leveraged transparency-enhancing technology in the financial sector, the fast-talking, stocky, bespectacled entrepreneur felt he could add value in agriculture, a field he initially considered "oblique."[3]

A digitalized ledger, Ramachandran explains, creates a clear and verifiable record of a meal's story in order to "build long-lasting trust and confidence in the food supply chain through a platform where everyone will be able to access transparent and reliable information on the origin, the journey, sustainability and the quality of their food."[4] No one party

Raja Ramachandran, CEO of ripe.io. *Credit: ripe.io.*

controls or manages a blockchain spreadsheet, and no one can alter or destroy the data. Such an information fabric, Ramachandran argues, allows farmers to collect, and obtain value for, vast quantities of data, including about soil conditions, farming procedures, animal welfare, and labor operations.

Ramachandran and Phil Harris, a banking colleague, in 2017 launched Ripe Technology, better known as ripe.io, in San Francisco's North Beach neighborhood, close to the Embarcadero. To prepare for the digitalization of the entire food supply chain, the pair began collecting studies about soils, seeds, and consumer tastes. To give farmers actionable data, they placed throughout numerous fields sophisticated soil-condition sensors that connected to smartphone apps. Moving along the supply chain, they helped distributors better predict crop-storage demands and manage warehouses and silos more intelligently. They enabled supermarkets—which operate on thin margins, with net profits of only 2.5 percent, making them among the least lucrative businesses in the United States—to predict future food-buying trends. To obtain insights on eaters, the entrepreneurs acquired FlavorWiki, a smaller start-up that measures consumer preferences associated with flavor, texture, and aroma. The pair also convinced several restaurants to share with their customers detailed farm-to-fork information, what they call the "food bundle," helping diners select meals according to their freshness, quality, and sustainability.

Ramchandran's initial motivation was to cut food waste, which he calls "a devastating problem."[5] One of blockchain's obvious uses is tracing where in the supply chain disease or spoilage occurs. Such information allows distributors to target food recalls and limit discarded food. It also enables farmers and processors to demonstrate compliance with government regulations as well as the sustainability goals of marketers.

A less recognized use is creating efficiencies that cut the margins enjoyed traditionally by middlemen. Eliminating the central controller of data, making that role transparent to all participants, reduces transaction costs. Something similar happens in the transportation industry, in which value created by increased supply-chain traceability benefits consumers.[6]

Blockchain also excites environmentalists. The reporting of emissions long has been difficult for regulators and activists to obtain and decipher. Sophisticated ledgers make discharges easier to track and verify, and they increase the accountability of polluters.

Blockchain's popularity should rise as sensors, drones, and artificial intelligence provide swelling amounts of data throughout the agricultural system. Ramachandran believes farmers increasingly will use that information to differentiate their products. "Blockchain," he says, "should help farmers receive monetization for what they do."[7] The ledger creates a verifiable trail to demonstrate that certain crops possess special attributes, be that sustainability, nonuse of genetically modified organisms (GMOs), or reduced use of pesticides. Ramachandran wants to move agriculture dramatically away from being a commodities business to one that delivers foods with distinct qualities.

The blockchain entrepreneur expects continued progress because "consumers want to get closer to farmers. The closer they get," he says, "the better they'll understand whether their food is safe and fresh."[8]

Yet Ramachandran admits blockchain "is not the end-all and be-all" to fix a broken food system. "It's not simply about tracking information," he says. "What is needed, and what ripe.io provides, is the analysis of that data in order to better understand what are the underlying factors that impact our food."[9] He maintains that scrutinizing blockchain data derived from a multitude of sensitive sensors allows all parties to identify the source of food-quality issues, be they from the soil, silo, warehouse, crop-transportation vehicle, or supermarket. Modern silo-based measuring devices, for instance, detect carbon dioxide concentrations three to five weeks earlier than traditional temperature monitors, allowing food distributors to take actions that prevent mold growth and pest infestation.

The ripe.io CEO also acknowledges that "the food business is vast" and advancing a farm-to-fork blockchain requires buy-in from numerous, and sometimes reluctant, players. Busy farmers, for instance, often do not have the time to collect detailed data about their fields and crop inputs; sensors and drones can do that work, but they impose a cost. Restaurant and market owners need to be convinced that their customers value knowing the story of their meals. Food processors and distributors must overcome their traditional view that all information should be proprietary and realize data's power to organize supply chains more effectively. Ramachandran asserts that as agricultural blockchains evolve, all these parties will discover how to monetize food-focused information.

To advance such buy-in, ripe.io in March 2019 launched a partnership with pork producers that monitors and evaluates their sustainability practices.[10] Through blockchain, says Ramachandran, associations such as the National Pork Board can help their members create and share trusted records that address food quality, traceability, and waste. Collaborations with such groups and co-ops also help ripe.io demonstrate to individual farmers the upsides of digitalization.

Ripe.io faces competition, even from tech giants such as IBM and Microsoft. The first blockchain-powered agricultural sale occurred in December 2017 when the Louis Dreyfus Company, a global merchant firm, arranged for the delivery of sixty thousand tons of soybeans to China. Since then, several start-ups have developed blockchains for separate commodities, such as seafood and coffee. Ramachandran acknowledges progress over the past few years but claims that "we've only moved the needle a little bit and many opportunities remain."[11]

The entrepreneur tries to differentiate his firm by providing information to consumers, who increasingly want to know about their food's quality but find such data difficult to locate and interpret. He declares, "We think there's a real strong future in matching the attributes of food you are directly consuming with how you think about and ultimately consume food." Ripe.io, Ramachandran says, gets "the farmer closer to the consumer and digitally [connects] those dots."[12]

CHAPTER 11

Lynette Kucsma and Emilio Sepulveda, Foodini—Printing 3D Meals

Remember *Star Trek*'s Replicator, the fantasy machine that synthesized meals on demand? Today, 3D food printers seem to be turning this science fiction into reality.

The US space agency, in fact, uses these modern appliances to feed astronauts on long-distance missions. Since space travelers clamor for something tasty to supplement their food pills and freeze-dried meals, the National Aeronautics and Space Administration in 2013 developed the Chef 3D, which "prints" a pizza with layered dough, tomato sauce, cheese, and spices. To cut waste and save space, astronauts rehydrate powdered ingredients and program the printer to create their pizzas' desired size, feel, appearance, and consistency. Russian cosmonauts on the International Space Station in 2019 printed cultured meat from the stem cells of a cow.

Industrial engineers began designing 3D printers in the 1980s, but the focus on food is relatively new. In what is called additive manufacturing, those early machines extruded layers of plastics to form toys, jewelry, and prosthetics. About a decade later, they laid down and laser-melted metal to make tools and precision parts. A market-research firm estimates 3D printing will grow by 26.4 percent annually between 2020 and 2024, becoming a $40 billion industry by 2024.[1]

Students at Cornell University in 2006 developed the first food printers, not surprisingly to make cookie dough and cheese crackers. Hershey's in 2014 worked with start-up 3D Systems to pattern various shapes from dark, milk, and white chocolates.

Lynette Kucsma—a cofounder of Natural Machines—argues that food-focused 3D printers will become as ubiquitous as microwave ovens, the last significant advancement in kitchen equipment. The Tappan Stove Company introduced the first microwave in 1955, but the large and expensive units (which would cost $12,500 in today's dollars) sold to only a few industrial bakeries. About a dozen years later, Raytheon presented a bulky countertop model, but consumers feared the emission of dangerous radiation and questioned the need for cumbersome machines in their kitchens. Sharp and Samsung subsequently marketed smaller and less expensive appliances, but it was not until 1997 that microwave ovens started becoming omnipresent.

"It took 30 years after microwaves went into the consumer market to get 90 percent market penetration," says Kucsma. "We see that being halved for a 3D food printer just because technology's advanced a lot, so we're a much more tech savvy audience, and plus we can build things out a lot faster. Our big vision is that in 10 to 15 years, 3D food printers will become a common kitchen appliance like an oven or a stove is in today's kitchen, [for] both professional and home kitchen use."[2]

Kucsma formed Natural Machines in 2012 with Emilio Sepulveda. "It all started as a conversation with a friend," she explains.[3] That friend owned a vegan bakery and wanted to sell sweet goods in other countries, but the costs of shipping from a central manufacturing facility proved to be prohibitive. The pair decided the better alternative was to create small and distributed food-manufacturing machines that "build" meals locally. Thus was born the Foodini.

When it came to printing food, initial efforts focused more on presentation than nutrition. Natural Machines, with offices in Barcelona and New York City, began assembling sweets and snacks, particularly elaborate meringues and chocolates. Using computerized recipes, their high-tech "chef" produced complex and colorful shapes made from sugar pastes that were not particularly tasty (or healthy) but added ornate highlights to the tops of wedding cakes and cookies.

Kucsma, who describes herself as health focused, soon shifted the firm's attention to fresh ingredients and more nutritious offerings. Maintaining that eating well improves a person physically, emotionally, and intellectually, she argues that Foodini allows individuals the flexibility to personalize individual meals and diets.

The Foodini, which measures almost ten inches high and weighs about twenty pounds, lays down layers of ingredients from five reusable stainless steel containers according to recipes devised or stored within its computer. The machine uses natural ingredients—be they in the form of a paste, gel, or liquid—to print a multitude of mixtures, including batters, jams, and doughs. To make ravioli, for instance, the machine first prints a layer of pasta, then adds a layer of filling, and then covers it with another layer of pasta. It is the same process a cook uses to make ravioli by hand, but Foodini automates it, leaves less mess in the kitchen, and enables creative shapes, varied ingredients, and different sizes.

One professor of agricultural economics compares a Foodini to a multicolor paper printer, which uses three base colors—red, green, and blue—to produce virtually every hue we see. "Similarly," he says, 3D food printers "eventually might use only a handful of base flavors that, when mixed in various combinations, create a myriad of familiar, and even unfamiliar, flavors and textures."[4]

Kucsma's meal-assembly machine comes with a variety of nozzles to deliver diverse textures. The smallest lays down a thin layer measuring 0.02 inch, which is hard for human chefs to obtain with knives or slicers.

Lynette Kucsma, cofounder of Natural Machines. *Credit: Natural Machines.*

The machine's touch screen (as well as any linked smartphone, tablet, or laptop) allows access to a library filled with thousands of recipes and shapes; select one and the printing begins. The newer models—which now appear in gourmet restaurants, smart kitchens, and bakeries—make ravioli, couscous, spinach quiche, crackers, pretzels, chocolate, chicken nuggets, and burgers.

Cooks can fill Foodini's stainless steel capsules with their own basic ingredients, prefilled cartridges, or packaged sets of ingredients for a specific recipe. On their way home from work, home chefs might stop at a supermarket's deli counter to pick up sets of fresh fillings and diverse doughs.

Lasers within some 3D appliances cook the food while it is being printed. Those concentrated beams, unlike microwaves, direct heat

Emilio Sepulveda, cofounder of Natural Machines. *Credit: Natural Machines.*

precisely and adjust to the needs of different ingredients. A blue laser, for instance, penetrates the food, while an infrared light browns the surface; combined, they achieve the desired temperature, texture, and cooking speed. Another option is to flash-cook or cool the mixtures within the printing chamber.

Foodini, according to Kucsma, offers five key benefits to chefs. The first is appearance, or the ability to wow the diner with intricate and creative shapes. "We eat with our eyes as much as our mouths," she explains. "Food presentation is very important." She notes that a Michelin-starred chef uses Foodini to dazzle his customers with a flower-shaped appetizer that resembles sea coral topped with caviar, sea urchins, hollandaise sauce, and a carrot foam.[5] Another cook displays a Foodini so her customers can marvel at their food being printed before their eyes. The devices can also produce pizza and other fast foods at sports arenas, festivals, university canteens, and theme parks; a fan orders from her app a pie, possibly in the shape and color of her favorite team's logo, and is notified when the meal is ready for pickup or delivery. KFC, in the fall of 2020, tested in Russia the 3D printing of chicken nuggets made from cultured chicken cells, plant material, and the company's iconic eleven herbs and spices.

Second, 3D printers perform prep work with speed and precision, slicing and dicing ingredients more efficiently than can human hands. Discussing a chef making quiche with numerous circles on top of one another, Kucsma asks, "Can he do it by hand? Probably. Would he get perfect circles? Probably not. But why should he spend his time doing 50 or 100 of those by hand when he can do it by machine and automate it and go do other things."[6]

Third, 3D printing customizes meals and tailors creations to the consumer's own biology.[7] Whereas restaurant servings of ravioli always include the same ingredients, the food printer creates a mixture that meets an eater's needed vitamins, level of hunger, or desired spiciness. Such customization helps those with dietary restrictions to control the ingredients in their meals. Kucsma foresees the appliance being linked to a user's personalized health information; it will, she says, know "from my wearable that I went on a 5km run and I'm low on vitamin D and iron, and can pump up the nutrients in my breakfast bar."[8]

Fourth, 3D printing reduces food waste. According to Kucsma, "ugly" but still highly edible leftovers can be "included into a more appetizing shape and thus kept in the food chain."[9] The printer, for instance, reprocesses nutritious cuts of fish that usually get left on the bone or discarded during the conventional food-preparation process. It also customizes portion sizes, printing only what the eater wants and nothing more.

Fifth, food printing encourages the use of fresh ingredients. Rather than buy prepackaged and shelf-stable foods filled with preservatives and too much salt, cooks can create healthier meals from scratch. Referencing spinach quiches printed in the shape of toy dinosaurs, Kucsma also suggests the 3D machine makes the preparing, and eating, of nutritious food more fun.

Despite these benefits, the food-printing industry faces several uncertainties, the first being whether consumers will eat something made by a printer. Will the machine-made product look, feel, and taste like what consumers expect? Kucsma counters that today's makers of processed foods sold in supermarkets already do essentially the same thing—pushing ingredients through machines, shaping them, and packaging the result. She suggests Foodini simply shrinks the large food processor down to a stylish appliance for a kitchen counter and gives consumers control of what to cook and eat.

The industry also must reduce the costs of specialized 3D printers, which remain pricey—ranging from $1,000 to $6,000—for small restaurants and individual homes. Foodini's sales price already matches the cost of existing cooking appliances in professional kitchens, and, according to Sepulveda, "when adoption and awareness grow, we will adapt to appeal to a broader market."[10]

Natural Machines initially targeted food professionals, including restaurant and hotel chefs, bakers, and deli workers. By the end of 2021, the startup plans to launch a new version with its patented laser-cooking technology for the home kitchen. "As Foodini is a completely new apparatus," says Sepulveda, "we found little value in using the traditional distribution chains, so we are focusing on marketing it ourselves."[11]

Another challenge is food-preparation time. Kucsma understands her product competes with a microwave zapping a prepared meal, but she claims each new version of Foodini prints quicker, with flat crackers already ready in twenty seconds and personalized pizzas in five minutes.

Kucsma and Sepulveda describe themselves as "unconventional explorers, chaotic creatives and crazy believers."[12] CNN in 2015 named

Kucsma one of seven "tech superheroes," and *Fortune* magazine says she "wants to sell the 21st century's version of the microwave."[13]

The entrepreneurs claim 3D printing enables chefs "to reinvent our culinary ways" and enables the customization of meals for shape, color, flavor, texture, and nutrients, thereby meeting an individual's wants and needs.[14] By creating meals on demand, 3D printing reduces food waste and inventories, and through sophisticated software, it empowers more people to cook their own fresh meals.

CHAPTER 12

Daphna Nissenbaum, TIPA—Cutting Plastic Packaging

"It's a product and a movement," boasts Daphna Nissenbaum of her compostable flexible packaging. "Plastic lasts forever," she complains, and it fills our landfills and litters our lands and oceans. To reduce such waste, the cofounder and CEO of TIPA creates packaging that breaks down into soil.[1]

Nissenbaum's aha moment came during an argument with her son, who had thrown away a plastic water bottle at school. She realized that other parents around the world probably were having similar conversations with their kids about the problems with plastics. She asked herself, "Okay, so what is the most natural way to do packaging? The first thing that came to mind was an apple. Because when I eat an apple [or an orange] the 'wrapping' is completely biodegradable."[2]

With that inspiration, she began researching the plastics market—a field that was entirely foreign to her. Nissenbaum began her career as a software engineer and became CEO of a research center on capital markets. She admits, "Packaging and plastics weren't really on my mind when it all started."[3] And when she turned to packaging, her first effort failed. After working for six months with independent biomaterial experts to design compostable beverage pouches to replace plastic bottles, she realized she was trying to solve the wrong problem, since water bottles can be recycled. The tougher challenge is flexible packaging, a $160 billion industry—which is expected to grow to $200 billion by 2025—that wraps vegetables, cheeses, breads, coffee beans, and snack bars.[4] Because of its chemical complexity, flexible plastic can rarely be recycled . . . and it lingers in the environment endlessly. Nissenbaum complains, "Every piece of plastic ever manufactured is still here."[5]

We produce globally each year more than 78 million metric tons of plastic packaging—thirteen times the weight of the Great Pyramid of Giza—and only 14 percent is recycled.[6] A growing amount—now at 9 million tons annually—escapes collection and flows into the oceans. The World Economic Forum estimates that by midcentury, plastics in the seas will outweigh fish.[7] Trash problems will grow worse as more people consume packaged foods and purchase more convenience items, including snack kits and takeout meals.

Nissenbaum recognized it was those wrappers that needed to be replaced. So, in 2010, she started TIPA with Tal Neuman to devise compostable alternatives to plastic packaging. TIPA's products are made to order. "We don't have 'on the shelf' items, because compostable packaging has a shelf-life," Nissenbaum explains, "and each run of packaging is customized to pack and preserve our clients' products."[8] Every specialized wrap, however, contains bio-based polymers made with renewable raw materials such as corn or sugarcane. While conventional single-use plastic packaging lingers for five hundred years or more in landfills, TIPA's products break down within six months in compost bins and are converted into rich fertilizer. Independent laboratories that run tests and issue compost certificates back up TIPA's sustainability claims.

Beyond keeping plastics out of landfills, Nissenbaum's broader ideal is a circular economy. "We can take all the plastic that we use today

Daphna Nissenbaum, cofounder and CEO of TIPA. *Credit: TIPA.*

that is on our shelves and in our homes and turn it into soil with compostable solutions," she says. "The big goal is to bring about change."[9]

Nissenbaum notes that conventional plastic packaging, with no viable end of life, contributes to global warming, the impacts of which she sees in record-breaking rains that flood cities throughout her native Israel. "Tackling climate change is a huge undertaking," she admits, "but I think it's something we can make a difference in."[10]

Indeed, like many others profiled in these pages, Nissenbaum sees her efforts in bold, some might even say grandiose, terms, saying that she wants to start "a packaging revolution." She appreciates that her products must offer the same durability, flexibility, transparency, and shelf life provided by conventional plastics, yet they need to eliminate waste and be fully compostable. "We can't entirely avoid packaging," she says, "but let's make it from the right materials."[11]

While Nissenbaum might be idealistic, she's also a realist, and she made an early decision to collaborate, rather than compete, with conventional plastic manufacturers. "TIPA is not against the plastic industry," she says; "we're working with it to create change." Recognizing the benefits of tapping into existing machinery and supply chains, she says, "If you want to be a mass market player, you have to be as smooth and easy to implement as possible. In this way, just replace one material with the other and not the entire infrastructure in the supply chain."[12] Nissenbaum also notes, "Revolutionary innovators are the companies that are able to think several steps ahead."[13]

Based in Israel but with offices in the United States and the United Kingdom, TIPA operates mainly in Europe, where its collaborators include supermarket chain Waitrose in the United Kingdom, cereal flakery De Halm in the Netherlands, and fashion designer Stella McCartney in the United Kingdom. Nissenbaum believes working with large, recognizable brands increases demand for packaging film, wrappers, pouches, and bags, and her ambition is to "collaborate more and more with massive players."[14] For her, the ability to scale up is a critical element of success.[15]

TIPA isn't the only company to make compostable packaging. In the United States, LOLIWARE produces an edible cup; looking like an ice cream cone, it holds either cold or room-temperature drinks. Wanting to replace plastic straws—almost two hundred billion are produced in the United States each year—the firm also markets a kelp-based tube that can be eaten after being used.

These companies are addressing a relatively new problem. Early in the twentieth century, food companies employed recyclable paper or cellophane, a flexible and compostable wrap made from plants. By mid-century, however, Dow chemists had turned to nonrenewable resources, mostly oil and natural gas, to develop polyvinylidene chloride, introducing Saran Wrap in 1949; the giant company switched to the less toxic polyethylene in 2004. This transparent film offered several advantages: it adheres to itself, blocks air and water, enables consumers to see what they are buying, and expands a product's shelf life by several days. As a result, use of plastic packaging accelerated, yet so did our throwaway culture and environmental contamination.

Synthetic plastics create more than trash; scientists slowly realized that petroleum-based packaging contains chemicals linked to obesity, cancer, and cardiovascular disease. Noting such problems, Trader Joe's in December 2018 announced it would make its packaging more sustainable and without harmful substances; Leonardo Trasande, a pediatrician and author of a book on hormone-disrupting chemicals, responded, "Avoiding the use of these chemicals of concern in packaging is a great step forward."[16]

Nissenbaum maintains that eschewing these chemicals requires new products. Of course, new products mean new business opportunities with innovations that reduce waste. "The role of technology is crucial," she says.[17]

In the male-dominated tech industry, Nissenbaum does not perceive being a female entrepreneur as a roadblock. "There are certainly differences in how women and men are treated in the startup industry," she explains. "Sometimes I feel them, but I choose not to see it as a challenge."[18] She finds the harder task to be turning an idea into a business. A key requirement is raising capital; although Nissenbaum does not discuss TIPA's finances, the company has attracted several venture capital firms, including Triodos Investment Management, Blue Horizon Ventures, and Greensoil Investments. She finds founding a company to be "a rollercoaster of emotions and experiences. . . . I'm always on my toes, because anything can happen tomorrow, especially in a start-up environment."[19]

Asked what inspires her work, Nissenbaum responds quickly: "My children's future."[20]

PART 3

CURTAIL POISONS

Farmers and gardeners have used pesticides for millennia. Sumerians about 4,500 years ago dusted sulfur on plants to prevent the growth of fungi. Over the years, growers have spread everything from natural pesticides made from chrysanthemums and tropical vegetables to toxic chemicals, including arsenic, mercury, and lead, to stop weeds and bugs. After World War II, scientists adapted for agricultural use numerous military synthetics, including DDT as an insecticide and triazine as an herbicide. Such chemical use has soared fiftyfold since 1950, annually accounting for 5.6 billion pounds worldwide.

To generations of farmers, chemicals seemed like the only plausible alternative to the backbreaking work of picking weeds and bugs off crops by hand. So it may be surprising that some of the most important recent technological developments in managing pests mark a return to that very method. But instead of human hands, technologists are taking advantage of robotic engineering, advanced sensors, and artificial intelligence to create giant weed-picking machines. Other innovators are working to spray unwanted plants precisely and to create pheromones that slow pest breeding. Taken as a whole, these inventions signal a significant move away from the synthetic pesticides that have been covering our produce and grains for decades.

These synthetics have had "catastrophic impacts" on human health and the environment, according to the United Nations.[1] One review found that "most studies on non-Hodgkin lymphoma and leukemia showed positive associations with pesticide exposure";[2] other researchers identified links between organophosphate insecticides and neurobehavioral ailments.[3]

Since the vast majority of pesticides drift, never reaching their targets, synthetic chemicals pollute waters, contaminate soils, destroy habitats, and reduce biodiversity.[4] The United Nations reports that the loss of species and habitats, much of it caused by industrial agriculture, poses as much danger to life on Earth as does climate change.[5] The US Department of Agriculture estimates that fifty million Americans drink from groundwater contaminated by pesticides and other agricultural chemicals.[6] The widespread use of synthetics also leads to resistant pests, raising fears of "superbugs" and prompting additional chemical applications.

Working with pesticides, moreover, is dangerous. The US Environmental Protection Agency calculates that almost 20,000 farmworkers each year suffer acute pesticide poisoning, and farmworkers' death rates are more than seven times those of other workers.[7] The damage is more severe in developing countries, where the United Nations estimates 200,000 people die—and another 25 million agricultural workers experience unintentional poisonings—each year from toxic exposure to pesticides.[8]

Farmers are not taking these risks without reason. Pesticides—including herbicides and insecticides—can prevent crop losses and save money. One study found that yields would fall by 10 percent without the use of synthetic chemicals. Yet farmers use these chemicals to tread water, not gain ground. In part because of pesticide resistance, pests and disease still destroy 30–40 percent of all crops, a figure that has remained constant since the 1940s, when ranchers and farmers began to deploy agrichemicals.[9]

Today, herbicides account for approximately 80 percent of pesticide use. Deployment of synthetic insecticides in the United States has actually fallen by more than half since 1980, largely because farmers began using transgenic Bt corn, which repels insects with natural ingredients. On the other hand, the use of Monsanto's Roundup herbicide increased fifteenfold since 1996, when the company introduced genetically modified seeds that withstand glyphosate.

Pesticides cost US farmers approximately $15 billion each year, a fivefold increase since 1960. Yet at the individual farm level, the chemicals remain relatively cheap. Add the facts that these synthetics face weak regulations and farmers avoid paying the costs of pollution, and growers have had little incentive to reduce chemical use. According to one study, "the average value of an acre of Florida tomatoes is about $14,000, while the average cost per acre for pesticides is about $750, or about 5 percent of the crop's value. Reducing pesticide costs by 20 percent, or $150, for example, provides virtually no potential economic reward compared with the perceived risk of change and the cash value of the crop."[10]

One of the key challenges for pesticide-alternative start-ups is to offer solutions that compete with the prices of their chemical counterparts. It can be hard to convince farmers facing tight margins that a huge robot with sophisticated sensors is the smart choice over the pesticides they have used for years. But there is progress on this front also. As advanced engineering becomes more widespread, costs drop, and with them, so can the use of dangerous chemicals.

CHAPTER 13

Sébastien Boyer and Thomas Palomares, FarmWise—Plucking Weeds Robotically

A San Francisco–based company—employing engineering graduates from the Massachusetts Institute of Technology, Stanford University, and Columbia University—offers autonomous machines that detect weeds and, mimicking a worker wielding a hoe, scoop the unwanted plants out by their roots—without using pesticides. FarmWise takes hundreds of thousands of pictures of different weeds and crops at various stages of development and under divergent growing conditions. With machine learning, the start-up creates images of what needs to be removed and what needs to be protected. Says cofounder and CEO Sébastien Boyer, "We basically had to leverage . . . recent advances in computation algorithms—algorithms similar to what Facebook and Google are using to recognize us in our pictures—to reach that level of consistency and repeatability across many different fields."[1]

After differentiating between sprouts that will become broccoli or other produce and those that will grow into invasive weeds, the robot deploys its pinpoint pluckers to remove unwelcome plants from the field. The precision process eliminates herbicide use, avoids hours of backbreaking manual labor, and helps resolve the farmworker shortages that plague growers. "A lot of farmers use herbicides and send crew afterward to pull the rest of the weeds," Boyer says. "This one process would replace that."[2]

Looking like an orange Zamboni or a boxy military tank, Farm-Wise's robot drives itself over the rows, picks targets, and extracts weeds by their roots. A single operator at a regional center can manage several machines in different fields at the same time, and the units work throughout the day and night.

FarmWise's robot. *Credit: FarmWise.*

Boyer and cofounder (and chief technology officer) Thomas Palomares met as undergraduates in their native France, where they promised to work together and do "something big." Boyer says, "We wanted to solve a large-scale, worldwide problem."[3] They maintained their dream and connection while Boyer completed graduate work at MIT, where he focused on machine learning and computer science, and Palomares, at

Sébastien Boyer (left) and Thomas Palomares (right), cofounders of FarmWise.
Credit: FarmWise.

Stanford, concentrated on management science and digital engineering. After graduation but still in their twenties, the pair turned toward farming, seeing opportunities to eliminate weeds without herbicides.

Boyer and Palomares launched FarmWise in 2016, believing that autonomous machines allow farmers to focus on individual plants rather than entire fields—and in the process increase productivity, reduce labor and herbicide costs, and protect the environment. The start-up occupies part of a series of two-story offices across the street from a shipping warehouse and next door to a dance studio, not far from San Francisco's central waterfront. Knowing that a robotic weeder enabled by artificial intelligence would need to integrate plant-perception algorithms, machine learning, and mechanical engineering, Boyer and Palomares initially had their computers view millions of images of a variety of plants, and they then imposed a precise set of rules for the machine to distinguish between weeds and crops.

The pair's grand mission is "to provide farmers with sustainable solutions to feed a growing world." Their tool is artificial intelligence. Their business model is to sell a service—weeding—rather than equipment—robots. They base their weeding charges on multiple variables, particularly comparable costs for labor and herbicides, but they claim their price always will "save growers money on their very first acres they are using the machines on."[4] They want to protect farmers from high up-front and ongoing maintenance costs. "We take care of our customers' weeding needs from A to Z, freeing them from the recruiting and maintenance hassles," explains Boyer. "Moreover, operating as a service enables us to offer the latest software and design updates to our customers."[5]

"We spent the first two years working closely with growers here in California going through a lot of iterations to take into account all of the requirements and constraints of fields," says Boyer. "Now we're providing a process to make it cheaper than ever to use less chemicals and farm organic fields."[6]

Cooperating with the Western Growers Association, FarmWise tested its robots in 2018 on lettuce and cauliflower fields across California's Salinas Valley and Santa Maria region. It has since proved the machines' success with other vegetables, including cabbage, carrots, broccoli, and celery, and it is expected to expand soon to strawberries and fruits as well as broader-acreage crops such as cotton and corn. Boyer predicts that as FarmWise multiplies its collection of pictures and data, the machines will adapt seamlessly to different crops.

"We have acquired very specific knowledge in the area of collaborative intelligent agricultural machines which allows us to build technologies general to various farming tasks,"[7] claims Boyer, who recently was listed on both *Forbes*'s 30 Under 30 and *MIT Technology Review*'s Innovators Under 35.

After proving their robot weeded well, Boyer and Palomares hit the entrepreneur's classic trial—to make their product in quantity. "We had a big challenge—getting everything ready for a huge scale-up of machine-making," Palomares says. "That meant all aspects of scaling—not just manufacturing, but support, shipping, hardware, and more as machines hit the fields."[8]

The San Francisco innovators agreed to partner with Detroit-based Roush Industries, a design and manufacturing firm with more than four thousand employees and a history in high-performance auto racing. Roush over the past decade also helped Google's Waymo project develop self-driving automobiles. Farming presents Roush with a new opportunity, according to CEO Evan Lyall: "From our end, we've seen it before in automotive, aerospace, and defense—the increasing use of automated vehicles. Now we're seeing it in agriculture."[9]

The partnership confronts competition. Deere & Company in 2017 purchased Blue River Technology, profiled in the next chapter, whose tractor-pulled robots spray tiny dots of herbicide on unwanted plants. A Swiss firm, ecoRobotix, offers autonomous weed-killing machines powered by renewable solar energy. According to a financial research firm, the market for agricultural robots will grow annually by 24 percent, rising from $1.01 billion in 2016 to $5.7 billion in 2024.[10]

Yet robotic farming faces skeptics. A few say the machines are too expensive and will drive farmers deeper into debt. Union organizers claim this "labor-saving technology" is nothing more than a clever means to eliminate agricultural jobs.[11] "This robotic farming future is not the unalloyed good the venture capitalists would have us believe," complains another critic; he claims modern technologies simply fashion "the future of farming without actual farmers or their knowledge."[12]

Yet FarmWise and Roush maintain they offer substantial benefits to farmers, consumers, and the environment. With plans grander than weeding, they predict their next robot will fertilize seedlings with the precise amount of needed nutrients. "Looking ahead," says Boyer, "our robots will increasingly act as specialized doctors for crops, monitoring individual health and adjusting targeted interventions according to a crop's individual needs."[13]

"We have to keep farming a profitable industry and make sure farmers thrive in this era," concludes Boyer. "We're providing cost efficient, affordable alternatives to chemicals and other harmful processes to farmers."[14] FarmWise also feels it's helping growers adapt to changes in US immigration policy that make it harder to obtain laborers; the start-up initially targets those growing high-labor crops, such as lettuce and peppers, who have a clear incentive to use machinery.

CHAPTER 14

Jorge Heraud and Lee Redden, Blue River Technology—Spraying Precisely

Like Sébastien Boyer and Thomas Palomares of FarmWise, Jorge Heraud and Lee Redden deploy robots to weed, but their approach is different and demonstrates the growing creativity associated with agricultural technologies. As graduate students at Stanford University, they envisioned spraying weeds precisely and, in the process, helping farmers save money and reduce environmental threats.

Heraud was raised in Peru, where his grandfather farmed. His father moved to California, pursued his entrepreneurial spirit, and convinced Jorge to focus on both engineering and business. "I remember summers where I spent the first part . . . helping out with farming," says Heraud, "and then came to California for the second half, to work on my dad's startup, helping with whatever I could."[1] Before heading to Stanford for an MBA, he spent fifteen years overseeing engineering and acquisitions at Trimble, a California-based software company.

Redden grew up in a Nebraska farming family, but he moved away to work at NASA's Johnson Space Center and Johns Hopkins University's Applied Physics Laboratory. With a PhD at Stanford in robotics and machine learning, Redden, who calls himself a "robotonist," displays a wide-eyed enthusiasm for large challenges and yearns for independence.

With a bit of funding from family and friends and a grant from the National Science Foundation, the pair in 2011 formed a firm that initially produced autonomous lawn mowers, but they attracted little interest. Like founders of Apple and other Silicon Valley start-ups, Heraud boasts that they initially "rented a little place that literally had a garage." As they moved into robotic weeding, they expanded into a larger space in Sunnyvale; the scrappy cofounder claims, "I bet we're the only startup with a tractor behind our offices."[2]

Heraud and Redden settled on a vision to make farming more sustainable through robotics and computer vision. They called their firm Blue River Technology to signify the clean waterways that would result from dramatically cutting agriculture's use of poisonous herbicides. Very appealing to the cofounders, says Heraud, "was the environmental impact we could have."[3]

Headquartered in Silicon Valley but not far from California's Central Valley, the pair turned their attention to lettuce, a major year-round crop that requires time-consuming, expensive, and labor-intensive thinning. Farmers plant lettuce seeds, the entrepreneurs learned, at high rates in order to ensure that enough plants emerge to make a uniform stand, yet growers must thin those crops in order to protect the best heads, provide spacing between plants, and obtain a uniform size for easy harvesting.

To automate that thinning process, Heraud and Redden developed cameras, sophisticated processors, algorithms, and precise sprayers that make more than five thousand decisions per minute about which small plants to keep and which to destroy. After squirting the unwanted lettuce heads with a small amount of herbicide, the robot takes additional pictures to verify its effectiveness.

The machine went through numerous prototypes, which Heraud

Lee Redden (left) and Jorge Heraud (right), cofounders of Blue River Technology.
Credit: Cindy Carpien for KAZU/Monterey.

and Redden carted to Salinas Valley for field tests and then back to their Silicon Valley lab for improvements. They officially launched Lettuce Bot in 2014, and by 2019 they were scanning and thinning 10 percent of all lettuce sold in the United States. That year they also won the prestigious AE50 Award from the American Society of Agricultural and Biological Engineers for having one of the most innovative engineering designs.

Heraud, however, wanted to do something more meaningful than just thin lettuce. To reduce chemical use and increase sustainability, he envisioned "mainstreaming artificial intelligence as an integral part of agriculture." The biggest problem facing farming, he says, "is how do we clean up our act, maintain scalability so that we can feed everybody—and even those who are to come—but do it in a sustainable way."[4]

"We could have gone in 1,500 different directions," admits Heraud, but Blue River's founders decided to concentrate on weeds, particularly those crowding cotton and soybeans. They called their new machine See & Spray.

Pulled behind a tractor like conventional spraying equipment, the forty-foot-wide robot uses thirty mounted cameras, supercomputing modules, and algorithms similar to facial recognition in order to distinguish crops from unwanted plants. The machine utilizes a database gleaned from taking and analyzing more than a million images, in different stages of growth, of pigweed, giant ragweed, cocklebur, and foxtail—to name a few common weeds—in order to recognize differences that might challenge the human eye. As the robot passes through a field at about seven miles per hour, its sharpshooting nozzles zap the identified weeds with a precise herbicide application about the size of a postage stamp that is tailored for the specific weed. (FarmWise, as discussed in the previous chapter, plucks rather than sprays unwanted plants.)

To recognize different crops and weeds requires deep learning, which didn't exist when the pair started Blue River but is advancing rapidly with significant improvements in software and hardware. "Traditionally, we would have designed an algorithm by using visual cues to identify the 10 or 20 parameters that best distinguish crops from weeds," says a Blue River vision engineer. "With deep learning, we can design a neural network that learns a million parameters to determine which are best for distinguishing crops from weeds. We do this by training the neural net with hundreds of thousands of examples from different farmers' fields."[5]

In addition to being smart, the robot and its precision equipment must be rugged enough to withstand a farm's heat and dust as well as the jostling through rutty fields. One Blue River engineer states, "We have to design components that maintain a camera's orientation to the crops to within a degree or two, all while taking a beating in day-to-day use. We found a lot of ways that didn't work before finding the one that did."[6]

See & Spray's precise applications save up to 90 percent compared with conventional herbicide spraying. Farmers enjoy reduced pesticide expenses, improved profitability, and fewer poisons leaching into groundwater or streams.

Blue River Technology, as a result, attracts notice. It's been listed among *Inc.* magazine's 25 Most Disruptive Companies, Fast Company's Most Innovative Companies, and CB Insights' AI 100 ranking of the most promising artificial intelligence start-ups in the world.

Yet Heraud and Redden recognize they must evolve—to revolutionize weed control in a world of growing pesticide resistance. Industrial agriculture's broadcast spraying of a limited number of herbicides has led to 250 species of resistant weeds that threaten to overrun farms. Blue River claims it gives growers a safer and more precise way to control such undesirable plants.

The entrepreneurs also want to change the scale of farming. Rather than treat an entire field the same way with a one-size-fits-all approach, Heraud claims robots can nurture specific plants with less cost to the farmer and less damage to the environment. Blue River's smart machines, he says, can "sense each individual plant, instantly determine everything about its health, structure and needs, and precisely apply the right amount of care."[7] With precision robots, he adds, farmers need fewer chemicals and fewer workers to perform the dirtiest and most dangerous jobs.

Although proud of their progress and prospects, the two recognized in 2017 that further success required turning "from a company that is a great engineering company to a company that also can manufacture reliable equipment, distribute it, support it, and grow the company in areas outside our expertise."[8] Their other option was to partner with an established corporation that could, so they accepted a $305 million buyout from Deere & Company. The giant firm committed to maintaining Heraud, Redden, and their team in Sunnyvale, yet time will tell if they retain their entrepreneurial spirit.

Now with additional resources, Heraud hopes Blue River will integrate advanced technology into all farm equipment, including planters, sprayers, and harvesters, and he wants his ultraprecise approach expanded to fertilizers, insecticides, and fungicides. To be profitable and sustainable, he asserts, agriculture must embrace robots.

CHAPTER 15

Irina Borodina, BioPhero— Messing with Pest Sex

Conventional insecticides kill virtually every insect in a field, including pollinating bees and natural predators of crop pests. Such poisons also harm farmers and birds, degrade biodiversity, and leave unhealthy residues on fruits and vegetables. A growing number of bugs, moreover, develop resistance to synthetic pesticides, forcing farmers to deploy more and more toxic chemicals.

Irina Borodina offers an alternative. Her approach—pheromone-based pest control—has been around for a few decades, but the founder of BioPhero devised ways to make pheromones more cheaply and to protect a wider variety of crops.

She focuses on sex—actually the blocking of copulation by cotton bollworms, diamondback moths, rice stem borers, fall armyworms, and other crop-destroying pests. Bordeaux winemakers first used such pest control in the early 1970s and, appropriately, called it *La confusion sexuelle* (English speakers say "mating disruption").

Confronted by the ravenous larvae of the moth genus *Cochylis*, the vineyard owners decided to try a bit of erotic trickery. They knew that female moths emit an airborne plume that contains a specific chemical blend designed to attract a mate of the same species. Horny males detect that pheromone at minimal concentrations of a few hundred molecules in a cubic centimeter of air and then track the scent for as far as thirty miles to locate the "calling" female. When impregnated, the female lays lots and lots of eggs, which develop into larvae that devour Bordeaux vines. To avoid such destruction, the vineyard owners decided to deceive the male.

The winemakers released a synthetic pheromone that closely mimicked—and therefore masked—the female insect's plume. Confused males could neither find nor impregnate females, which, to the delight of the winemakers, cut the rate of successful mating and collapsed the insect infestation.

Other farmers subsequently directed such mating disruption toward pests damaging apples, pears, and other high-value crops. The process proved effective and provided numerous complementary benefits. Covert pheromones, for instance, confuse only the targeted species, protecting beneficial insects, pollinators, and birds. Since the mating disruptors contain no toxins, evaporate over time, and do not persist in fields or on crops, the US Environmental Protection Agency judges them to be among the most environmentally friendly means to exterminate pest invasions. Compared with conventional and toxic insecticides, pheromones do not harm farmworkers, and, finally, insects do not develop resistance to such sexual confusion.

Yet mating disruption has suffered challenges. First, the processes to develop and produce deceiving pheromones have relied on expensive and specialized chemicals as starting materials. As a result, sex inhibitors have been limited to 0.1 percent of the world's fields that grow fragile produce (approximately one million acres) and have not been deployed on corn, soybeans, rice, and other row crops that represent more than one-third of global farmlands (about one hundred million acres). Participating growers, moreover, have had to release the misleading plumes by

Irina Borodina, founder of BioPhero. *Credit: BioPhero.*

hand, which is time-consuming and costly. Third, because conventional mating disruption targets a single destructive insect, farm fields with multiple pests have required a more complex approach, often referred to as integrated pest management, that includes various pheromones, soil management, and crop rotations.

Borodina wants to eliminate those limitations. After obtaining a chemical engineering degree in her native Lithuania, she earned a doctorate in biotechnology from the Technical University of Denmark (DTU), where she became senior scientist at the Novo Nordisk Foundation Center for Biosustainability. She heard at a conference about the expense of producing pheromones, so she set out to create a high-tech means to produce the substances cheaply. She founded BioPhero in 2016 as a technology spin-off of her DTU laboratory.

"It's all about metabolic engineering," Borodina declares. She developed a biotechnology-based alternative to conventional pheromone production. In a scientific paper, she explains, "This required reconstruction of synthetic biochemical pathways towards pheromones in yeast, extensive engineering of the yeast host to improve the flux towards the products, and optimization of fermentation processes."[1]

Borodina's process, which she claims will transform pest management, begins with identifying the specialized enzymes that insects naturally use to create pheromones. She then expresses an identical pheromone into yeast, which comes from a family of cells widely used in food production. This engineered yeast ferments in large tanks, similarly to those used to brew beer, but the output is pheromones rather than alcohol. "This biotechnological process," the scientist adds, "is the same as the one used to produce enzymes for food and washing powder, and insulin for diabetes."[2] For farmers and consumers concerned about sustainability, her yeast feedstocks are renewable and the pheromones are biodegradable.

Borodina works with enzyme-laden yeast because of its safety and productivity; she says the process "can produce sex pheromones that are identical to the ones that the female insects secrete to attract males."[3] Pheromone-laden yeast also grows in industrial-scale fermenters, which means large quantities of sex-retarding agent can be produced relatively cheaply.[4] When growers release their low-cost tonics in the fields, male insects become confused and fail to find mates; the unfertilized females cannot lay eggs, and no larvae emerge to damage crops.[5]

Borodina first concentrated her mass-produced pheromones on moths, particularly the corn borer, whose larvae destroy grains. She then turned her attention to the cotton bollworm attacking Brazilian soy fields and the fall armyworm devastating corn crops in Africa.

Borodina wants natural pest disruptors to be used widely and works with other firms to improve distribution—and application. Growers or their contractors in the past placed by hand throughout the fields polymer capsules that control the release of droplets of pheromone; another approach trapped insects on glue boards containing pheromone bait. In contrast, farmers can spread BioPhero's cheaper product far and wide with crop dusters or tractor-towed sprayers, both common machines that producers already utilize.

Borodina has written forty-five peer-reviewed articles and a dozen patent applications. The European Union in 2019 selected her for its EU Prize for Women Innovators. To concentrate on scientific research rather than business matters, she recently hired a CEO with finance and sales experience, who established BioPhero offices in Copenhagen and New York City.

Borodina is not the only scientist trying to produce pheromones on a large scale. Agenor Mafra-Neto founded ISCA Technologies in Riverside, California, and deploys natural chemicals called semiochemicals to manipulate the behavior of insect pests. He began to develop sustainable means to protect crops after a close friend died from exposure to toxic pesticides.

The agricultural pheromones market was valued at $1.9 billion in 2017, but business analysts expect it to grow by almost 16 percent annually to reach $6.2 billion by 2025.[6] In contrast, yearly sales of chemical insecticides surpass $17 billion, suggesting to Borodina that alternative approaches have the potential to expand substantially. According to Grand View Research, "growing awareness regarding food security and concerns over the ill effects of synthetic crop protection chemicals are expected to drive the [pheromone pest-control] market."[7]

Borodina claims her product is already less expensive than sprayed synthetic chemicals and can be used at a large scale on row crops.[8] Part of that advantage comes from having to apply pheromones only once per year, during the flight phase of the insect, whereas insecticides generally have to be sprayed several times per year. She also argues that new advances in biotechnology will further reduce BioPhero's production costs.

Borodina sees pressure for change coming both from consumers who are increasingly wary of pesticides and from the pesticide industry, whose high development costs limit the pipeline of new synthetic insect killers. Worried about health and environmental damage, regulators also restrict the chemical insecticide options available to growers; on the other hand, a cumbersome regulatory system makes it hard to obtain approvals for environmentally beneficial pesticides, thus prolonging the use of more harmful chemicals. Farmers want safer biological alternatives, says BioPhero: "The future of pest control is not killing insects. It's disrupting their mating patterns."[9]

Borodina calls pheromones a smarter way to control pests and sees them as the next frontier in crop management. To her mind, protecting the environment and human health from pesticides comes down to science.[10] The pheromone entrepreneur declares, "At BioPhero, we believe the answer is innovation."[11]

PART 4

NOURISH PLANTS

Plants need food, but they do not eat the way we do. They make their own nutrients from light, air, and water through the remarkable chemical process of photosynthesis. As we all remember from biology class, plants absorb and store energy from sunlight (or artificial light will do) with a green pigment in their leaves called chlorophyll. That energy converts carbon dioxide from the air and water from the soil to form molecules of starch and sugar, which the plants, often with the help of microbes, use to build tissue and grow.

We do not "feed" plants, and they do not "eat" the soil. But roots do absorb from the soil various minerals—including iron, nitrogen, magnesium, potassium, and calcium—that crops need to be healthy. While farmers long have added manures or fertilizers, hoping to supplement these nutrients, soil health is becoming a science. Start-ups use sophisticated sensors and chemical analyses to gauge plants' nutrient levels, they produce natural probiotics that nourish specific microbes for different plants, and they devise new means to bring nitrogen to individual crops. Their sensors and solar-powered controls also provide just the right amount of water needed throughout a plant's life.

A growing number of innovators focus on microorganisms, the multitudinous mixture of tiny creatures living in soils that convert various elements into a form plants can use to grow and produce food for us. That mixture—including bacteria, fungi, nematodes, and protozoa—decomposes organic residues and recycles soil nutrients, and several beneficial microbes create symbiotic associations with plant roots and even block pests, parasites, and diseases. More of these invisible creatures exist within a teaspoon of soil than there are humans on Earth. Also in the soil are larger earthworms, mites, nematodes, and ants, whose burrowing aerates the ground and makes it easier for water to penetrate and reach roots.

Nourishing the soil and its inhabitants, then, is critical to life on Earth. Unfortunately, industrial agriculture strip-mines our rich soils. Its giant plows grind up nutrients and expose the land to erosion, and its mechanical sprayers douse fields with chemicals that kill useful microorganisms.[1] Some farm scientists calculate we have lost about half the soil's nutrients over the past fifty years from industrialized and intensive agriculture. Rebuilding the soil takes a long time, with some estimates of only a single percentage point gain in five years.[2]

As synthetics dislodge the soil's organic matter, they increase the demand for additional and more toxic chemicals. Ammonia-based fertilizers, moreover, leach substantial quantities of nitrates into surface water and groundwater as well as into the atmosphere as nitrous oxide, a potent greenhouse gas.

A few Big Ag firms develop products that boost the efficiency of their synthetic fertilizers. Koch Agronomic Services, as noted before, markets nitrification inhibitors as chemical additives to the firm's ammonia-based fertilizers in order to slow nitrogen losses within soils. Other large corporations produce chemical supplements that block the natural enzyme urease and thereby reduce the evaporation of ammonia.[3]

Numerous academics, moreover, research ways to make agriculture more efficient. The international RIPE project, housed at the University of Illinois and funded by the Bill & Melinda Gates Foundation, tries to supercharge natural photosynthesis and increase crop yields. Scientists model the 170-step photosynthesis process, which utilizes the sun's energy for plant growth, and speed up the activation of Rubisco, the key enzyme that converts carbon dioxide into sugars that sustain crops.

Yet most innovation in soil health comes from entrepreneurs devising technologies—including genomic testing, drones, computerized controls, and sensors—that help increase the productivity of soils. But what these start-ups are really selling is information—precise data and analytic tools that give farmers a clearer picture of their fields and what's needed to make them more nourishing.

CHAPTER 16

Diane Wu and Poornima Parameswaran, Trace Genomics— Mapping Soils

Just south of San Francisco International Airport and on the northern edge of Silicon Valley, Diane Wu and Poornima Parameswaran map soils and the health of microbes living within them. Agricultural lands can be incredible reservoirs of biodiversity, but many fields' microorganisms have been drenched and destroyed for decades by chemical fertilizers, pesticides, and plowing.

Wu and Parameswaran's start-up develops DNA kits for soils, not unlike genetic tests such as 23andMe that provide insights into a human's ancestry, health, and traits. Farmers collect fifteen grams of soil, equal to about three teaspoonfuls, and then pay the scientists to analyze the samples for bacteria, minerals, viruses, and fungi that affect how well crops will grow. By digitalizing that vast amount of data and comparing it through proprietary algorithms with thousands of other soil samples, Wu and Parameswaran give farmers insights into what seeds to plant, what nutrients to feed the soil's microbes, and how to avoid diseases that lurk in the dirt. "Making these kinds of decisions for the growing season has been an art," argues Parameswaran. "Now we want to make it a science."[1]

Healthy soil, a farmer's greatest asset, is rich with microorganisms, yet the biological properties within that dense and diverse world are virtually unknown. Parameswaran estimates we understand a little about 10 percent of the soil's microbes, and nothing about the rest. She and Wu seek to reveal that world to us by taking advantage of powerful microscopes, increased computing capacity, and new gene-sequencing technologies.

The two met during graduate school at Stanford University, where Wu focused on biochemistry, software engineering, and machine learning while Parameswaran concentrated on molecular biology. They both worked in the lab of Professor Andrew Fire, a Nobel Prize–winning geneticist who nurtured a passion among his students for innovation. Colleagues comment on the pair's sense of humor but call them the nation's smartest soil scientists as well as masters of genomics technology and big data.

Wu and Parameswaran formed their firm, Trace Genomics, in 2015 with the goal of creating "the most comprehensive agricultural soils database in the United States."[2] They began collecting samples from farms growing lettuce, strawberries, and almonds but quickly expanded their collections to specialty crops, such as potatoes and apples, as well as to row crops, such as soybeans and corn.

The start-up each month adds to its library of soil samples from thousands of acres, making its insights more valuable to growers and agronomists. Trace Genomics keeps information from individual farms confidential, but its data bank identifies commonalities among soils and uncovers lessons from the practices of diverse cultivators. Wu explains, "Detecting patterns was impossible just years ago, but we're now helping farmers increase profits by measuring what's in their soils."[3]

The scientists take advantage of great advances in computer analytics, in which processing power has risen exponentially and changed the ways we communicate, travel, and heal. "We digitize the living soil,"

Diane Wu (left) and Poornima Parameswaran (right), cofounders of Trace Genomics.
Credit: Trace Genomics.

say Wu and Parameswaran. The firm applies "proprietary soil DNA extraction and sequencing to index and quantify millions of microbes in your soil."[4]

We then "decode the soil data," say the Trace Genomics cofounders. "Harnessing machine learning, we conduct high-speed, cost-efficient data analysis to compare against a large and growing set of soil data."[5] Their core market is farmers, who use their information to make better decisions about their crops. Jerry Dove, an Iowa farmer, attests, "It's good to see this new data, so we can put values on how we are making improvements."[6]

The start-up's founders also tackle broader environmental problems. Their belowground insights reveal the damage done by industrial agriculture's multidecade war on beneficial microbes and highlight how that harm can be reversed. And as climate change threatens crops and accelerates diseases and pest invasions, they use genomics analysis to calculate whether individual fields could benefit from drought- or heat-resistant plants and resilient microbes.

With this information in hand, Trace Genomics identifies biological seed coatings that stimulate soil microbes to release nitrogen and other nutrients that benefit crop roots. (Chapter 18 profiles an innovator creating such coatings.) "It's a new field," says Parameswaran, "exciting because you are creating new tools and applications."[7]

Although based on natural ingredients, those applications have raised the ire of some advocates of organic and regenerative agriculture. "The microbial communities that exist in animal guts and in the soil have evolved over eons," asserts author Tom Philpott, the food and agriculture correspondent for *Mother Jones* magazine. "I suspect that diverse diets and crop rotations—not lab-grown potions—are key to engendering healthy biomes, both within our bodies and in the dirt."[8]

Despite this criticism, the coatings allow farmers to vastly reduce their spraying of expensive synthetic fertilizers that pollute waterways. Trace Genomics' data and analytics also help cultivators reduce disease. Soybean growers, for instance, fear a fungus that causes sudden death syndrome and attacks plants so late in the growing season that little can be done to treat the disease and save the crop. By utilizing Trace's soil intelligence, however, farmers can detect traces of the fungus early, in time to protect their plants proactively.

In addition to its detractors, the firm faces numerous competitors and challenges. Other biotechnology firms—such as Benson Hill in St. Louis, Missouri, and AgBiome in Durham, North Carolina—utilize cloud-based biology and genomics to offer advanced screenings for the agricultural sector, and Alphabet's X lab, the former Google division that launched the Waymo self-driving car, has developed a four-wheel rover-like prototype, what it calls a "plant buggy," that studies soils and environmental factors with a mix of cameras and sensors. To show unique value to farmers, many of whom distrust new technologies, Wu and Parameswaran recognize they must raise more capital, lower the costs of soil studies, and reduce the analysis time from days to seconds.

Although neither founder has worked the land, both claim a family relationship, with Wu's grandfather a third-generation produce grower and Parameswaran's uncle the owner of a local fishery. Being newcomers themselves to farming, the pair initially talked a lot with cultivators, learning their perspectives and concerns. They split their time between Silicon Valley and Salinas Valley, the productive agricultural region located a bit south of the San Francisco Bay Area. "We were just sponges ready to soak it all in," says Parameswaran. "For the most part we listened."[9]

Trace Genomics' founders raised venture capital funds and created a diverse team, but their passion remains science. So, in 2019, they hired Dan Vradenburg as chief executive officer to manage operations and expand consumer outreach, particularly to row-crop farmers in the Midwest, while Wu focuses on engineering and Parameswaran on product development. Vradenburg spent thirty-five years building Wilbur-Ellis, a multibillion-dollar firm that distributes animal feed, seeds, fertilizer, and machinery.

Funding the firm was no easy task, especially since women occupy only 7 percent of leadership positions within the nation's top venture capital firms. When they started, Wu and Parameswaran were the only, or two of the few, women at ag-tech conferences. "It's hard not to notice," says Parameswaran, "but if anything, it makes you more determined. It's an opportunity to show the world that you're like the other gender; there's nothing different."[10] Since Trace Genomics' founding in 2015, however, the pair have witnessed a lot more diversity. Parameswaran says, "It's exciting to see more women in general, minority women, and we are seeing this trend in ag-tech." Noting that farmers want straightforward analytics and advice to overcome problems, she says, "Whether it comes from a woman's mouth or man's mouth, it doesn't matter. I think it's all about your major focus being about helping them solve their needs."[11]

Wu and Parameswaran are taking the guesswork out of farming and providing actionable information for growers and their agronomists. As the speed of topsoil evaluation quickens, the two innovators give farmers a more holistic and science-based view of their operations and options. The results will be increased yields, healthier soils, and ultimately more sustainable agriculture.

CHAPTER 17

Eric Taipale, Sentera—
Analyzing Fields from Above

Another way to map working lands is from the sky. Drones—small, unmanned aircraft—became the rage on farms and ranges during the early 2010s. Initial aerial imagery provided fresh perspectives on fields and livestock, but growers and ranchers quickly became dissatisfied with simple pictures. Those early views had to be uploaded cumbersomely to a personal computer located in an office miles away from the field, and farmers increasingly complained that by the time the processed images alerted them to problems, it was too late to apply remedies.

Even drone company executives admit their early offerings were rudimentary. "To be frank," says Eric Taipale of Sentera, "showing growers and agronomists pictures was enough and it was easy for us to start with."[1] Yet Taipale acknowledges that farmers want more than images; they need "actionable intelligence" that increases their output and profits.[2] He also appreciates that start-ups can't confine themselves to single problems, such as mapping excessive soil moisture, but must expand to tackle diverse issues.

Taipale—with an engineering background and twenty years of experience with aerospace and sensors—shifted Sentera's focus from simple pictures to data analysis. "We pivoted hard into machine learning and data sciences to take the data off those sensors and help the customers make use of it," he says. "We count plants, find weeds and predict yields, look at disease and nutrition. A couple of years ago the customers would have had to determine that from the imagery. We now provide the value-added data analysis."[3]

Taipale enjoys displaying a McKinsey chart that shows how various industries use digitalization. Ranking at the bottom is agriculture, which he interprets as an opening for businesses such as his. The fact that most farms haven't yet taken advantage of low-cost sensors on drones, satellites, and tractors doesn't mean they won't. It helps that computing costs to process vast quantities of information are plunging.[4]

Speaking in the language of an engineer, Sentera's CEO suggests growers are in the manufacturing business, and they want to eliminate as many production uncertainties as possible. Weather and market prices may be out of their control, but new technologies allow them to understand where their soil is compacted, the levels of nutrients and microbes, the best seeds for different soils, moisture concentrations, and the locations of weeds and disease. In the past, farmers could do little more than sample small segments of their fields and extrapolate from previous trends. Taipale says, "We want to take uncertainty out of a farmer's decision loop."[5]

The entrepreneur feels that agriculture is "the last frontier of manufacturing digitalization," which he believes opens opportunities for innovation. "I can't think of another industry," he says, "where data and analytics like the type we provide have the potential to generate so much additional economic value."[6]

As an example, Sentera's CEO describes sending a drone above a large field to identify unwanted plants. The company's proprietary software converts data from the aircraft's sensors into a detailed centimeter-by-centimeter map, with shades of green showing weed concentrations.

Eric Taipale, CEO of Sentera. *Credit: Sentera.*

Sentera automatically downloads that map to a high-precision self-driving tractor that then applies herbicide only where it is needed, cutting chemical costs by three-quarters.

Taipale developed apps for handheld devices and customized how they present and interpret data and maps. Such applications must overcome obstacles, not just identify them. "Now customers expect us to deliver highly precise analytics that automatically detect problems and give prescriptive action immediately," the entrepreneur says. "We are not just providing 500 pictures, but providing an answer."[7]

Similarly to other innovators profiled in these pages, Taipale focuses on precision, obtaining specific data that are difficult, if not impossible, to obtain through industrialized or preindustrialized methods of farming. Whereas Trace Genomics evaluates microbes within the soil, Sentera focuses on the crop's leaves, using cameras to detect bug infestations, diseases, and other threats. Pinpointing such problems helps growers reduce yield-cutting outbreaks, target their soil additives, and lower costs.

Like Jorge Heraud and Lee Redden of Blue River Technologies, whose robots spray weeds precisely, Taipale enables farmers to reduce their use of synthetics, which often leach into water supplies. By microtargeting threats, both start-ups dramatically cut the use of poisonous (and costly) agrochemicals.

Based in Richfield, Minnesota, a bit east of Minneapolis–St. Paul International Airport, Sentera focuses on corn, soybeans, and cotton, but Taipale sees opportunities with strawberries in California, almonds in Australia, and citrus in Florida. The firm already deploys its digital crop "scouting" and real-time analytics in sixty-two countries.

Sentera is not alone. The firm is riding a wave of growth in drone use in numerous arenas, including military, human relief, and disaster management. Aircraft costs in recent years have fallen dramatically—heavy-duty drones now cost just a few hundred dollars—while the sophistication of sensors and speed of computerized analytics have risen remarkably. Although the use of drones once involved steep learning curves, modern models are relatively easy to program and control, even for those without tech savvy.

Unmanned aircraft seem particularly apt for monitoring soil health, identifying disease, spreading seeds, planning irrigation, spot-spraying fertilizers and pesticides, estimating crop yields, and monitoring livestock.[8] According to the Bank of America, the US agricultural drone industry between 2015 and 2025 will create an additional one hundred thousand jobs and generate $82 billion in economic activity; farm applications, moreover, soon will account for 80 percent of the commercial market for such aircraft.[9]

As with many technologies, a primary motivation for using drones is to cut costs. One grower saved 17 percent of his pea crop after a drone spotted an area of his field with aphids, a sap-sucking insect, and then sprayed that targeted region. By avoiding the expense of a duster or tractor dousing the entire farm with toxic pesticides, this focused approach, according to the farmer, "more than paid for the drone."[10]

Another cultivator saved tens of thousands of dollars when a drone discovered an outbreak of fungus in a small corner of her large field. In the past, she might have found that problem if she had happened to be walking in that specific area, and she would have responded by spraying expensive fungicide across the entire farm, putting workers and nearby streams at risk. In contrast, drone mapping gave her real-time diagnosis of her whole acreage, and aerial spot spraying allowed her to treat only the affected zone.

Drones with thermal cameras helped yet another grower identify irrigation issues. On farms with miles and miles of water lines, leaks are hard to find, but aerial maps locate them as well as display sections of the field receiving too much or too little water. To avoid damaging delicate crops, the drone's analytics also show the best planting plans that account for the land's natural contours as well as minimize runoff or water pooling.

Drone proponents also argue that unmanned aircraft mean less on-the-ground work that often is dangerous, time intensive, and expensive. And they suggest drones relieve farm labor shortages and allow quicker reactions to storms and droughts associated with climate change.

Sentera's multiple sensors clip onto a variety of drones. One imager with red edge capability calculates changes in vegetation's reflectance, identifying plants that overheat during photosynthesis. Thermal sensors build high-resolution, true-temperature maps that highlight which insects, weeds, and diseases thrive in particular soils. Other instruments provide electro-optical and infrared readings that measure acidity and nutrient levels, while another offers detailed GPS location tracking. The analytics package, moreover, snaps orthophotos that measure true distances after accounting for topographical relief, lens distortion, and camera tilt.

Multispectral (four bands of light) and hyperspectral (forty bands of light) images help create normalized difference vegetation index (NDVI) maps that reveal plants under stress, the growth of weeds, and the prevalence of pests. Although plants within a field may look the same in visual light (the combination of red, green, and blue primary colors), the leaves of healthier crops reflect more light in the near-infrared range. NDVI maps, therefore, reveal dehydrated or strained vegetation.

While sensors on drones provide very high resolution and complex images, Sentera's software incorporates measurements from satellites and ground-based sensors that can help evaluate the soil's moisture, nutrient content, and compaction. Adding in data about what was planted and when, what was applied and in what amounts, and what were the results from prior-year efforts, says Taipale, helps "to move the grower closer to optimal."[11]

Sentera's CEO thinks these technologies will become more and more common as measurement and computing costs continue to fall.[12] To win over skeptical farmers, he knows his start-up must do a better job of converting the complexities of engineering into practical guidance. According to Taipale, "We are an adjunct data source that provides the grower and their trusted crop advisor with better data, faster and with better coverage to help them make optimal decisions for their farmlands."[13]

Another strategy for growth is to work not only with individual farmers but also with the large corporations that buy their products. In 2017, Taipale approached the giant beer producer Anheuser-Busch InBev (AB InBev) and asked what problems its hundreds of grain suppliers faced. High on the list was lodging, the bending over of a plant's stem that complicates harvesting and lowers yields. "AB InBev asked us if we could build a tool that could detect what percentage of barley had lodged in a field, and could we do this at a pretty significant scale," says Taipale. The partnership, launched in March 2020, initially focused on ten thousand AB InBev growers in North America who possess the networking infrastructure needed to capture and process data from drones.[14]

AB InBev views the partnership as integral to its sustainability goals, which call for cutting carbon dioxide emissions across its supply chain by 25 percent by 2025. Taipale feels that the arrangement helps farmers embrace digital technology and understand that "sustainability and economic success go hand in hand."[15] He says, "It's a 'chicken or the egg' kind of scenario. Until there are enough users, there isn't a market for these kinds of analytical insights and until there are enough analytical insights, there aren't enough supportive users. AB InBev, due to the sheer size of their grower network, can bring both."[16]

Other companies collect and analyze data to improve farm outcomes, yet Sentera's number one competitor, says Taipale, is "nothing," when farmers decide against adopting any new technology. "The biggest hurdle is getting that first acre," he states, "working with a customer to convince them to give our technology a try instead of doing what they've always done."[17]

In addition to cutting carbon, digitalization can curtail other types of waste. "No one wants to pump water that's not needed," says Taipale. "No one wants to overtill the soil. No one wants to waste money on unnecessary fertilizers and herbicides."[18] While companies like Anheuser-Busch InBev increasingly face pressure to meet sustainability goals, spending extra to do so can be a hard sell in corporate boardrooms. If sensor and computing technologies actually save the company money, it becomes a very different discussion.

Taipale is no stranger to designing tech for large corporations, having worked as chief technology officer for FourthWing Sensors and spent several years at Lockheed Martin Corporation, including as lead engineer for its Autonomous and Unmanned Systems business unit. His passion is engineering, merging an array of technologies, including drones, measuring instruments, and computers. "I couldn't ask for a better job," he declares. And while he has no background in farming, he takes pleasure in supporting growers and ranchers. "The knowledge provided by these technologies," he says, "will help a farm family whose ancestors worked the fields for one hundred years to hold on to those lands and make a decent living for another hundred."[19]

Taipale also believes Sentera tackles our generation's greatest challenges. "We can help produce more food to feed an expanding population," he says, "and we can do all of this while we deliver massive environmental benefits to our planet. It's not a bad way to make a living!"[20]

CHAPTER 18

Ron Hovsepian, Indigo Ag— Providing Probiotics to the Soil

Far from any farm, Ron Hovsepian's three-story brick-and-glass head-quarters lies in a low-rise industrial area north of Boston, not far from the Bunker Hill Monument. Neither a farmer nor a scientist, Hovsepian views himself as a serial entrepreneur. He led the venture firm, Flagship Pioneering, that spun out Indigo Ag in 2013, joined the start-up's board in 2019, and took over the reins there in 2020. He previously worked as president of Intralinks, which creates platforms for document sharing, as well as CEO of Novell, which provides software that networks personal computers.

In the fast-paced world of start-ups, where individuals can rise and fall without warning, Hovsepian unexpectedly took over as Indigo's CEO from David Perry, who had led the company for almost six years and had raised $1.2 billion. Indigo would not explain the executive shift, but the company's carefully crafted press release credited Perry with "building a transformational system to de-commoditize agriculture and drawing attention to the role agriculture can play in addressing climate change." It called him "a motivational leader that led Indigo Ag through its first evolutionary phase."[1]

By late 2020, Indigo Ag employed about 1,100 staff and was valued at more than $3.5 billion, making it one of the best-funded ag-tech start-ups. CNBC annually ranks the top fifty disruptive not-yet-public companies, which in the past included Airbnb with lodging and WeWork with office space; in 2019, the business network featured Indigo Ag at the top of that list for its efforts to transform how food is grown, distributed, and sold.

Hovsepian, ever the moneyman, continues to work as Flagship Pioneering's executive partner, and he says he is "committed to a path toward profitability [for Indigo Ag] with an increased focus on customer-centricity and operational excellence." He suggests business partnerships are key to efforts to make agricultural ecosystems "more efficient and beneficial for the planet."[2]

That language is pretty standard in the ag-tech world, but Indigo can boast some significant innovations. The most obvious is the science itself. Since decades of industrial farming—with its persistent plowing and chemical spraying—have stripped beneficial microbes from soils, Indigo sifts through seed and plant libraries to identify which microbes were in or on heritage plants but are missing from their modern counterparts. The firm's scientists also travel to farms around the world seeking out microorganisms associated with thriving plants as compared with their weaker neighbors. The start-up then isolates the bacteria and fungi inside healthy crops that allow them to flourish in different soils; to withstand harsh conditions, such as droughts or pest infestations; and to sequester carbon dioxide, the major contributor to climate change.

Indigo coats seeds with non-GMO microbes that then live within the plant's tissue, improving crop yields with fewer water and chemical inputs. The company sells the coatings to farmers as either a spray or a powder. Rather than staying in the soil or on the plants' leaves, some of the microbes become part of the plants' internal biology, making them

Indigo markets its technology platform as a way to bring farmers and buyers together.
Credit: iStock by Getty Images/Cleardesign1.

motivated to help maintain the crops' health. Unlike the periodic spraying of synthetic fertilizers, these nimble microorganisms also respond quickly to environmental changes.

"Indigo's collection of beneficial, proprietary microbes," the company boasts, "has the capacity to protect crops from nearly every abiotic stress imaginable—from extreme temperatures and water scarcity, to nutrient-deficient soils. By helping shield plants from tough conditions and enhancing their use of resources such as water, Indigo microbes have the potential to improve yields, increase farm revenue, and ultimately reduce the need for fertilizers."[3]

Indigo points to numerous studies showing how beneficial bacteria inoculate plants against pathogens. One analysis found that symbiotic microbes delivered to eggplant roots reduced wilt by 70 percent, while another discovered that such inoculations activated plant defenses in sugarcane and rice. According to David Montgomery of the University of Washington, "applications of the right kinds of bacteria prepare plants to repel pathogens through teeing up their defenses the way commensal [or "friendly"] gut bacteria help keep the [human] immune system poised to repel pathogens."[4]

Indigo initially focused its seed treatments on cotton, which becomes stressed without sufficient moisture. That stress is likely to grow as fresh water becomes increasingly scarce and costly; in part because climate change amplifies droughts, water use on farms is expected to soar by 50 percent by 2050. So the company started adding microbes to cotton-seeds, creating the equivalent of probiotics. As a result, company officials claim to see real differences in stem diameters and root mass, the latter of which is an especially good predictor of a plant's ability to manage water stress. Indigo calculates that its initial microbial treatments—on five thousand cotton farms totaling one million acres—achieved 10–14 percent increases in yield, and Hovsepian anticipates that additional experience will raise that gain to the 25 percent range. After cotton, the firm developed microbial seed coatings for four row crops—wheat, soybeans, rice, and corn—and it is researching applications for coffee, high-value nuts, fruits, and vegetables. To amp up international sales, the company opened offices in Argentina, Brazil, Australia, and India, and its technology in 2020 spread to twenty-five thousand farmers working four million acres. In a bold claim, Indigo Ag estimates microbes will come to replace half of today's chemical fertilizers and 90 percent of synthetic fungicides and insecticides.

Indigo's newer coatings also encourage desirable traits in plants. With the appropriate microbes, for instance, soybeans increase their oils and seeds, becoming more appetizing to specific livestock.

Indigo's second innovation is the way it sells its product. To entice wary farmers, the start-up charges little for its coated seeds but then takes a 33 percent share of any additional self-reported value they create for the farm. As a result, Indigo profits only when its microbes perform well. Stetson Hogue, whose family has been growing cotton for generations outside Lubbock, Texas, likes this approach: "We were looking for any way to find an edge."[5] In a 2017 blog post, David Perry argued that his firm and farmers become partners, saying, "The business of farming needs to be economically attractive, bringing in new farmers, in order to sustain and expand the farming community."[6]

Indigo's signature product is the seed coating, but the firm is trying to make bigger changes in how crops are valued and how farmers are compensated. The start-up's third advance is to move agriculture away from being a commodity business, in which undifferentiated crops are purchased solely on the basis of price and convenience.[7] Indigo maintains plants are not equal. One farmer's wheat might be high in protein, which bread millers prefer, while his neighbor's wheat contains less protein and is favored by beer brewers. Rather than mix both types in a conventional grain elevator, Indigo's crop-tracking system allows farmers to profit from what is singular about their plants while buyers obtain what they most need. In providing farmers with new revenue for their current practices, one member of Indigo's team explains, they are "creating value from nothing, or sometimes actually just stopping value from being destroyed."[8]

Hovsepian recognizes consumers' growing interest in knowing where their food comes from and how it is produced. To meet the demand for more transparency, his company sells a digital platform that connects growers and buyers. The start-up envisions farmers earning extra if their crops provide specific traits consumers want, such as capturing carbon or avoiding synthetic chemicals. It's a system that values quality over quantity, which the CEO believes could help with today's farms' razor-thin profit margins.

Since distinguishing crops requires data, and agriculture generally lacks digitalization, Indigo created what journalist Charlie Mitchell calls "a global network of sensors, drones, and satellites that send one trillion data points *per day* back to Indigo's Boston headquarters for analysis."[9] The firm also conducts detailed research about crop attributes with 125 large farmers who tend about a million acres.

Indigo is not alone in seeing the value of differentiating crops. Barry Parkin, with the giant candy company Mars, argues that "the commodity era as we've known it is over." It's time, he says, "to radically simplify our supply chains and commit to buying from known and verified suppliers."[10]

Indigo grew its ability to gather data in late 2018 by purchasing TellusLabs and its geospatial satellite technology. "In the same way that Google maps can tell you anytime a new Starbucks appears on the corner," said Perry not long afterward, "we want to have that same level of information about our food supply. As an example, we now know every field that corn and soy is growing on in North or South America. We can estimate the yields on those and update that on a daily basis."[11] Farmers, with access to that information, can make better decisions about when to plant and what they'll earn at harvest.

Indigo is also getting involved in the carbon market. In the summer of 2019, the start-up announced its Terraton Initiative, which pays cultivators to cut one trillion tons of carbon dioxide from the atmosphere. Indigo offers initial participants $15 per metric ton sequestered, and it then sells those carbon reductions to companies seeking greenhouse-gas offsets. One company, FedEx, invested in Indigo to ensure it can purchase carbon credits to reach its sustainability goals, and Indigo raised another $360 million in August 2020 for this carbon exchange. The carbon credit market is poised to grow, and by 2020 farmers controlling 21 million acres had expressed interest in this voluntary exchange.[12]

Indigo uses ground sensors and satellite images to help participating farmers assess their soil's health and carbon content. The start-up

acknowledges that carbon markets work only if we can measure sequestration, and it estimates that the average farmer deploying innovative
ag tech can sequester two to three tons of carbon per acre per year. (On
a 500-acre farm and with carbon priced at $15 per ton, for example, a
carbon-reducing grower could earn an extra $22,500 annually.)

Indigo monitors each farm's methods of sequestering carbon and
measures changes to the amount of carbon in its soil. Like Raja Ramachandran with ripe.io, Hovsepian uses distributed ledger technology
to verify and secure the carbon reduction calculations. Growers benefit
by selling certified credits on the carbon offset market (or, in California, within the state-mandated cap-and-trade program), while Indigo
acquires a new customer for its seed coatings and gets paid as the carbon credit middleman. Indigo estimated farmers would sign up three
million acres in Terraton's first year, but the company exceeded that
goal sixfold.

Sequestering carbon dioxide in the soil, said Perry, "is the most hope-
ful potential solution that I know of to address climate change. It's the
only [solution] being talked about today that is on the scale of the problem."[13] Indigo also argues that carbon-rich soils increase yields, resist
droughts and flooding, and require fewer chemical treatments.

The innovative start-up, as noted earlier, offers a mix of services,
including seed treatments, informed planting advice, and carbon markets. One company executive says that "no one technology or algorithm" will create a winning model, "but the combination, a systems
approach, is what will take us there."[14] Yet that very combination—
based on gathering enormous quantities of data—makes some industry analysts nervous. Journalist Charlie Mitchell notes that Indigo Ag's
approach comes with risks: "it means buying into a supply chain that
Indigo owns and controls completely." That concern about a single
company obtaining an information monopoly is valid, yet the engage-
ment of a variety of data-based firms—including giants such as IBM

and upstarts such as ripe.io (profiled in chapter 10)—suggests that, at least for the moment, competition will advance innovation as well as protect both farmers and consumers.

CHAPTER 19

Karsten Temme and Alvin Tamsir, Pivot Bio—Feeding Nitrogen to Crop Roots

Karsten Temme and Alvin Tamsir believe in the power of microbes. The pair met at the University of California, San Francisco, while studying biology and computer science, and in 2010, they moved across the bay to Berkeley, where they founded Pivot Bio. Growing tiny soil-based bacteria, the company is working to blunt the damage caused by synthetic fertilizers.

Temme admits fertilizers increase crop production, but, he argues, they do so only temporarily and at a steep price. "Farmers know that maybe half of the chemical fertilizer they apply is actually used by the crop," he says. "Chemical fertilizers are volatile—they don't stay put."[1] Rains often wash away these costly, nitrogen-saturated additives, which accumulate in approximately five hundred "dead zones" around the world, the largest in the Gulf of Mexico. Algae thrive on the runoff's nitrogen and suck up so much of the water's oxygen that no aquatic life survives; in addition, these toxic blooms of decomposed algae, sometimes called hypoxic zones, kill marine mammals and birds. "That's a lot of money to waste," says one farmer, and it "has a substantial environmental impact."[2]

Karsten Temme, cofounder and CEO of Pivot Bio. *Credit: Pivot Bio.*

Synthetic fertilizers also stimulate the release of nitrous oxide, a greenhouse gas 265 times more potent than carbon dioxide and one that lasts 120 years in the atmosphere. The factories that manufacture these chemicals, moreover, release as much greenhouse gas as does the heating and cooking within all houses in the United States.

Noting these threats, Temme asks, "How can we make it so that crops don't need fertilizer?" His answer: "Create a more sustainable and efficient food system."[3]

Through their university studies, Temme and Tamsir learned that some bacteria suck up nitrogen from the air and convert it into nutrients that plants can take up through their roots and turn into proteins.

Alvin Tamsir, cofounder and CSO of Pivot Bio. *Credit: Pivot Bio.*

The plant, in return, provides sugar to the hungry microbes, using photosynthesis to transform sunlight into energy.

The researchers also discovered that synthetic fertilizers outcompete natural nitrogen-fixers, causing them to "turn off." Since fixing nitrogen demands a lot of energy, soil microbes drenched in synthetic fertilizers over the past century simply stop this process and use their resources on other activities; the genes associated with nitrogen fixing have become dormant.

Temme and Tamsir offer a means to bring the crop microbiome out of hibernation and re-enable it to utilize atmospheric nitrogen. They claim their technology, which they call "computationally guided microbe remodeling," will "replace all nitrogen fertilizer with microbes that adhere to the crop's root system and spoon-feed the crop each day."[4]

"We go in and break the wiring in the microbe that connects their ability to sense nitrogen in the soil to their decision to become a source of nitrogen for the farmer," says Temme. "The goal is less fertilizer required."[5]

Temme and Tamsir initially sought to combine genetically a microbe and a plant. Says Temme, they had hoped to "make a self-fertilizing plant by taking the whole DNA program and putting it into a plant."[6] Yet that process proved too difficult—and consumers tend to fear genetically modified crops—so Pivot Bio essentially pivoted to a crop additive that tweaks the dormant parts of a microbe's genome so the plant again fixes nitrogen. The entrepreneurs call their microbe-focused product PROVEN.

The young scientists first decided to focus their specially fermented soil bacteria on corn, although the company now remodels microbes for wheat, sorghum, soybeans, and rice. To find the best bacteria, they spent several years examining hundreds of soil samples from throughout the Midwest's Corn Belt. "We were looking to single out microbes that had a mutualistic relationship with corn out of the billions that live in the soil," said Temme. "We call it 'shovelomics' because it starts and ends with the soil beneath our feet."[7]

Using genomic sequencing, they "reawaken" the nitrogen-producing abilities of naturally occurring microbes. "We found a way for microbes to rediscover their own potential to take nitrogen from the air and make it available to plants," explains Temme.[8] "Our scientists fine-tune that molecular machine by editing the microbe's genetic instructions to produce nitrogen daily, as the corn plant needs it. The microbes go through rigorous testing in the lab and greenhouse to ensure they can perform under modern agricultural field conditions and once we've identified the most robust microbes, we advance the best microbe candidates to the field."[9]

Put another way, Pivot Bio adjusts the genetic switch that had been preventing microbes from fixing nitrogen while they lived in chemically fertilized soils. With that signal changed, the firm's biofertilizer prompts natural microbes to again nourish crops even when fields contain synthetics. Temme compares the process to rearranging furniture within a room or to breaking a negative feedback loop within the genome.

Temme goes out of his way to say that Pivot Bio does not genetically modify its microbes, since the start-up avoids transferring genes from one species to another. Instead—much as with the work of Rachel Haurwitz, the genome pioneer profiled in chapter 21—the firm edits the genetic material already within soil bacteria. Through biological genetics, its researchers adjust the single signal in the bacteria's DNA so the microbes produce nitrogen that adheres to a plant's roots.

Temme claims that Pivot Bio's crop-nutrition additive "provides sustainable solutions for farmers and improves the health of the planet through scientific innovation."[10] He further describes his mission as devising nitrogen-producing microbes that make "agriculture an environmental force for good."[11]

Farm trials conducted by six land-grant universities calculated that Pivot Bio's nitrogen-fixing microbes increased the average corn yield per acre by six bushels, or about 5 percent;[12] Temme and Tamsir maintain such yields will rise as the technology matures. One farmer observed that Pivot Bio's treated plants enjoy "bigger stalk diameters, a little more ear-fill on the tips, and some deeper kernels," all of which "add up to yield."[13] The start-up calculates that if 35 percent of US corn production utilized this innovative approach, nitrous-oxide emissions each year would fall by nearly twenty thousand metric tons, which equals the greenhouse-gas emissions from 1.3 million automobiles driving 14.8 billion miles.

Like so many working in ag-tech, neither Temme nor Tamsir grew up on a farm. Temme says the closest he got was on summer vacations when his family drove from his home in Wyoming to his grandparents' house in Michigan and he gazed out the car window at vast midwestern fields. Yet he claims, "We're entirely motivated to help farmers keep their family businesses alive and improve their economics, especially in places that are still developing around the world."[14]

The scientists, despite their good intentions and positive initial outcomes, admit they face skepticism from farmers, particularly those who have grown used to the ease offered by synthetic nitrogen fertilizers. Other growers distrust new technologies or feel disconnected from Silicon Valley entrepreneurs. "I wish startups understood our margins and our economic climate better," complains Bryan Tomm, who participated in a Pivot Bio trial on his 2,500-acre farm in southeastern Illinois. "I think farmers live in a pretty isolated world when it comes to the economy, and the farm economy works very differently than traditional Silicon Valley might work. It seems like some startups come out with a product that can help me make $20 more per acre, but they charge $21 per acre for the product. Everyone is after the dollar. The exciting technologies are the ones that really bring value and increase profitability for the farm."[15]

Temme appreciates that Pivot Bio must show farmers true cost reductions and yield increases. He admits that the soil microbiome, with billions of bacteria interacting, is still largely unknown. While skeptics doubt scientists can sort out such complexity, Pivot Bio claims it focuses on the few microbes that prove best at fixing nitrogen. Temme explains that whereas nitrogen in synthetic fertilizers is volatile and can be carried away in extreme weather, preventing 40–60 percent of chemical nitrogen from reaching the targeted plants, his microbes provide "more predictable yields and a more dependable nitrogen supply."[16] Pivot Bio's product, moreover, prompts microbes to emit nitrogen gradually throughout the growing season, eliminating the need for multiple applications of synthetic fertilizer. One satisfied farmer says, "With this product, I can set it and forget it."[17]

For growers to adopt PROVEN, of course, the company needs to increase production capacity and compete with other new enterprises, including Joyn Bio, a $100 million partnership between the chemical giant Bayer and Ginkgo Bioworks, a Boston-based synthetic-biology company. Temme and Tamsir also must answer other questions: Will adding microbes to the soil, for instance, cause harmful side effects? Will a single bacterium work cooperatively with multiple plants? What will happen to crop yields after several years of microbial applications?

Temme recognizes the tall order associated with disrupting the $155 billion synthetic-fertilizer industry, and he says that "creating a next-gen fertilizer is the hardest problem in agriculture."[18] Yet he foresees Pivot Bio producing enough enhanced microbes to meet every crop's nitrogen needs.

Temme maintains that as beneficial microorganisms increase farmers' yields and profits, growers will shift away from synthetic chemicals that threaten their soils and the planet. "As we help the microbes get better at producing nitrogen," he predicts, "it'll end up being a full solution."[19]

CHAPTER 20

Tony Alvarez, WaterBit— Watering Precisely

Farms and ranches guzzle water. Agricultural irrigation accounts for almost 80 percent of the water taken from US rivers, lakes, and aquifers, far more than any other human activity. Unfortunately, we waste about 60 percent of that resource, equal to almost 400 trillion gallons.[1] Tony Alvarez believes he can profit by cutting that inefficiency.

Alvarez in 2020 became the board advisor for WaterBit, a Silicon Valley–based start-up whose sensors and autonomous controls make irrigation more precise. The company's field-based monitors—each smaller than a lunch box but equipped with a long-range radio and powered by a solar-electric cell—evaluate soil conditions and plant stresses, and its water-flow regulators seek to "achieve consistent soil moisture levels automatically and remotely—improving crop quality and yield, while saving water and optimizing labor."[2]

Alvarez, a self-described tech entrepreneur, held leadership roles with SunEdison Semiconductor, Advanced Analogic Technologies, and Leadis Technology. "Although this is an unprecedented time for our country and the world, farming can't stop and growers need to continue to look for ways to make their operations more efficient," he says. "In fact, it's probably more important than ever for farmers to get the most out of all available resources."[3]

Irrigation long has been a labor-intensive and inexact process. Growers typically spend countless hours driving around their different fields to check on moisture levels and manually turn on and off hundreds of water valves and sprinklers. On large operations spread out over miles, such labor costs become substantial.

Irrigation is also inefficient. A few farmers work in regions with sufficient rainfall, but most lands require watering. Surface irrigation spreads huge volumes over fields, either by submerging crops or through furrows or trenches on either side of a row of plants. Farmers use surface irrigation to grow rice and corn; it is most suitable for clay (but not sandy) soils. Sprinklers, which can be permanent or movable, deliver water in different pressures and varying droplet sizes depending upon their guns and nozzles; this method covers wide areas, but 30 percent of the water is lost to evaporation, and large globules tend to bombard pollinators and flowers.

Less wasteful irrigation for most crops comes from a trickle or drip system, which began millennia ago with clay pipes or pots containing small holes. A big efficiency gain occurred in 1965 when Netafim introduced micro-irrigation technology on a kibbutz in Israel's Negev Desert; the company laid thin plastic pipes that dripped small, targeted amounts of water to nourish individual plants. This approach starves weeds by directing water only to the crop, but drip systems tend to be expensive to install and maintain.

WaterBit digitizes drip or precision irrigation, claiming that its high-tech monitors, analytics, and variable rate valve controls automatically deliver the right amount of water at the right place at the right time. The company installs sensors throughout a farm because watering needs differ according to various soil types, topography, plant demands, and exposure to sunlight. Each device takes electrical readings that calculate

Tony Alvarez, board advisor of WaterBit. *Credit: WaterBit.*

soil moisture tension, which determines how much water the soil holds and needs. A reading of zero suggests dried-out dirt that's been baked in the sun; another at 52 reveals the consistency of muck.

Those monitors also measure each plant's stage of development, water flows, and the status of valves and irrigation pipes. They even consider weather forecasts to ensure farmers do not water their fields before a rainstorm. As a result, WaterBit's autonomous irrigation system evaluates the crop's stresses, compares them with previous plant performances, and uses an algorithm to provide the amount of water needed for best results.

The sensors connect to a single gateway, which sends all that granular data to the cloud and provides analysis in real time to growers via their computers or mobile devices. The preprogrammed sensors also communicate with the valves, turning them on and off as needed to deliver the precise amount of water to specific plants in specific parts of the farm—all without the grower leaving her pickup truck, office, or vacation home. Taking advantage of Alvarez's role with Solaria, a global solar systems company, WaterBit's technology is powered by the sun; as a result, the farmer does not need to change batteries or string electrical wires, and she avoids the labor and time associated with manipulating spigots by hand. Alvarez argues that his company's seamless system "gives what the plants tell it to give." It is "like a thermostat, only for water."[4]

WaterBit, formed during California's 2016 drought, performed one of its first tests at the vineyards owned by its chairman, T. J. Rodgers, who founded and long led Cypress Semiconductor. Having obtained numerous patents associated with silicon chips and photovoltaic cells, Rodgers helped create WaterBit's computerized irrigation system.

To measure each vine's water needs, Rodgers initially consulted with professors from the School of Viticulture and Enology at the University of California, Davis. He tried monitoring leaf water potential, which indicates the plant's moisture status, but obtaining such measurements proved to be labor-intensive, and the insertion of sensors into a vine's water-carrying tissue wounded the plant. He then turned to calculating the soil's volumetric water content (VWC) around the vine's root zone, but available sensors were too expensive and inaccurate. So he invested in WaterBit, a start-up that promised to produce a low-cost VWC sensor.

To test the device, the company placed its new instruments throughout Rodgers's vineyard—named Clos de la Tech, in Woodside, California—at depths of 2, 10, and 14–22 inches. The system triggers short drips of water when the VWC falls to a programmed target level. When the deeper probes report moisture readings, the irrigation system automatically turns off to ensure water is not lost to percolation past the roots.

Initial results proved encouraging as WaterBit's system cut vineyard water usage by up to 52 percent compared with conventional drip emitters. To explain the results, Rodgers says, "I hypothesized that the extra water used in the control irrigations probably percolated through the rootzones, doing no good at all."[5] Also saved were greenhouse-gas emissions associated with truck trips needed to monitor the fields.

Grapes from the traditional and WaterBit approaches were fermented and barrel-aged separately, and to Rodgers's delight, crop yields proved to be identical, and blind taste tests revealed similar wine quality. The experimental barrels, in fact, qualified for inclusion in Clos de la Tech's reserve wine, sold at $130 per bottle.

Another WaterBit customer, Devine Organics, found that the precision irrigation doubled its yield of leafy green vegetables. Water usage fell by 6 percent, and greenhouse-gas emissions declined by almost 10 percent.

Alvarez and Rodgers face substantial competition. Google's executive chairman, Eric Schmidt, invests in CropX, an Israel-based start-up whose wireless sensors collect soil data that inform the application of irrigation water. KETOS, headquartered in Silicon Valley, offers real-time water-quality intelligence that helps farmers detect impurities in surface water and groundwater. Moleaer, based in Southern California, provides oxygenated nanobubbles that give water and nutrients to crops; one of the company's early projects was to help the National Aeronautics and Space Administration grow food in the microgravity of space. The cofounder of Hortau, another California-based precision-irrigation firm, says, "Agriculture is currently in the early stage of a technological revolution in terms of innovating and leveraging the information-age tools to grow more food with less inputs and strain on the environment."[6]

High-tech watering companies attract investors, raising $670 million in private equity in 2015 alone.[7] According to business researcher Hexa Reports, the emerging industry's sales could reach $43.5 billion by 2025.

Alvarez and Rodgers feel they are bringing farming into the Internet of Things (IoT), a system of interrelated sensing and computing devices that transfer data without human intervention. Rodgers calls this future the Internet of Plants.

Sophisticated irrigation is particularly important in this era of climate change. As water becomes more scarce—because of less rain or snow or because a plant's demands increase with escalating temperatures and droughts—US growers need to be far more efficient or they will be forced to abandon crops or farmlands and try to rebuild in cooler climes farther north.

These stresses will only intensify as farmers face ever-tighter margins and higher water costs. Taking advantage of advances in sensors and computation, says Alvarez, smart irrigation can reduce resource waste and increase sustainability.

PART 5

CUT CARBON

Agriculture is a victim as well as a cause of climate change. Growers increasingly confront unexpected floods that bury fields, late freezes that destroy buds, and hotter temperatures that advance crop-destroying pests and pathogens. For example, growers of tart cherries around Northern Michigan's Grand Traverse Bay for decades enjoyed long, cold winters and slow, cool springs. Because of global warming, however, the region no longer reliably freezes, warmer temperatures arrive too early for blossoms to bloom, and more violent spring storms pummel trees with strong hail and high winds. Such climate changes caused two total crop failures in the past decade, and according to the Northwest Michigan Horticulture Research Center, "the stress is becoming too much for many of the growers."[1]

On the other hand, agricultural practices prompt global warming. Tillage exposes organic matter that decays and emits carbon dioxide. Livestock belch methane, a more powerful greenhouse gas. Synthetic fertilizers prompt the release of nitrous oxide, an even more potent pollutant. When farmers slash trees to grow more food for more animals, they eliminate one of the best retainers of carbon.

Advocates of organic and regenerative agriculture argue that the key to reversing this damage is a return to preindustrial practices—avoiding synthetic pesticides and fertilizers and instead adding nutrients with an artful combination of manure, reduced tillage, and cover crops, particularly during the winter months. Certainly these practices have become more popular, but even their advocates admit that the additional farm management tempers their adoption. Organic sales in 2018 surpassed $52 billion,[2] more than double what they were a decade before, and a few of the largest companies—including Amazon-owned Whole Foods Market and General Mills–controlled Cascadian Farm—have joined the trend. Yet, while organic farms have grown to more than 5 million acres, rising by 20 percent between 2011 and 2018, they still account for less than 1 percent of the country's total croplands.[3]

Like so much with agriculture, organic farming presents trade-offs. Environmentalists hail its reliance on compost rather than petrochemical fertilizers, as well as its spraying of plant- or bacteria-derived agents instead of toxic pesticides. Yet by trucking tons of compost and plowing regularly to keep down weeds, some organic farmers burn more fossil fuels than do their conventional colleagues. As one analyst observes, "There's no requirement that an organic dairy farm have a lower climate footprint than a conventional farm."[4]

Another trade-off affects grass-fed cattle. Most proponents of regenerative agriculture oppose concentrated feedlots that fatten animals with grains and antibiotics; they support the regular moving of livestock to new pastures, where they eat the tips of grass but allow roots to expand, thereby preserving erosion-curtailing ground cover that pulls carbon into the soil. Yet grass-fed cattle belch up more methane, and they spend more months burping because they take longer to reach their slaughter weight—emitting about 20 percent more greenhouse gases than do their grain-fed counterparts. A Harvard University report found that grass-fed operations also are less efficient than concentrated feedlots, requiring more cattle—and more pollution—to achieve the same meat production levels.[5]

Farmers, as noted earlier, tend to complain that regenerative-ag practices are complex, costly, and labor-intensive. Despite hundreds of philanthropy-supported reports and conferences, as well as admonitions from high-profile chefs and authors, cover crops shield only 2 percent of US farmland. Even the World Resources Institute, which favors organic and regenerative agriculture, admits that preindustrial practices do little to battle climate change: "We consider large estimates of climate change mitigation potential for agricultural soil carbon to be unrealistic. Many published claims of massive potential do not address [the] scientific and practical challenges."[6]

Ag-tech entrepreneurs believe they offer an alternative to both industrial and preindustrial farming approaches, a third path that is the better answer to mitigating climate change. While innovation and regeneration advocates differ, particularly when it comes to the wisdom and potential of technology, they seek to make farmers and ranchers part of the greenhouse-gas solution. Both camps would benefit by recognizing their similar goals and searching for shared strategies. To highlight that potential, the profiles in the following chapters describe how high-tech entrepreneurs regenerate soils and cut greenhouse-gas emissions.

This book, as stated before, profiles entrepreneurs using private-sector capital and markets to disrupt conventional agriculture. It is worth noting, therefore, that this section contains one slightly different profile—Lee DeHaan at The Land Institute, who breeds deep-rooted perennial grains that sequester carbon in the soil. His approach is innovative and significant, but most of his development money comes from government research grants and charitable contributions rather than equity investments. Still, DeHaan's sketch is included because perennial crops could be disruptive and because the private sector finances the grain's marketing.

Not included, however, are other creative and significant innovations—including Neil Black's efforts with California Bioenergy (Cal-Bio) to obtain methane from the manure-laden lagoons of concentrated animal feeding operations. Although Black is an engaging innovator—having edited the liberal *Nation* magazine and obtained an MBA from Harvard—CalBio's business model depends upon state legislation that requires reductions in cattle methane emissions as well as upon large state grants that help feedlot owners finance the expensive digesters that capture methane to generate electricity.

Profiled in the chapters that follow are five other entrepreneurs attracting private equity for ag tech that reduces greenhouse gases. Geneticists edit genes so that crops better absorb and retain carbon. An aquaculturalist creates feed supplements that cut cattle's methane-laden burps. A biomedical engineer assembles a sustainable-food platform that makes it easier to bring climate-friendly foods to markets. A chemist devises food whose production process absorbs carbon dioxide. While none could be confused with a regenerative farmer, each shares the goal of making agriculture a solution, rather than a cause, of climate change.

CHAPTER 21

Rachel Haurwitz, Caribou Biosciences— Editing Genes

Farmers and ranchers long have manipulated plants and animals to their liking. Early *Homo sapiens* domesticated wild grasses to create barley and einkorn wheat, and modern breeders mix crops or creatures to obtain desired traits, such as sweeter corn or meatier cows. As one researcher put it, "There isn't a single crop that I know of in your produce aisle that is not *drastically* modified from what is out there in the wild."[1]

To accelerate such manipulations, scientists in the 1930s began to expose organisms' genes to radiation, carcinogens, or high temperatures in order to disrupt their DNA, and a few of the mutant varieties demonstrated useful traits that breeders tried to exploit. Perhaps the most noted results of such mutagenesis are Rio Red grapefruit and the barley hops used to ferment beer. Yet despite such high-profile foods, irradiation's scrambling of chromosomes takes a decade or more to develop a desired characteristic.

The green revolution, beginning in the 1960s, accelerated these efforts with hybrid crops—created by cross-pollinating two different varieties of a plant to contain the preferred qualities from each parent, such as thicker stalks and stronger roots that allow corn plants to stand straighter and better tolerate mechanical harvesting. Farmers, unfortunately, discovered that a hybrid's second generation fails to produce as many kernels as the first, with yields typically falling by one-third. No longer motivated to save a portion of their seeds for the next planting, growers had to buy new hybrids every spring from large seed-supply corporations.

Breeders even developed meatier chickens that grow faster yet eat less feed. By concentrating on a few chosen varieties, however, they reduced the diversity of poultry, and they increasingly fattened these quick-growing birds in large, mechanized warehouses that generate tremendous amounts of pollution.

In the mid-1990s, Monsanto went further to engineer foreign DNA into the genomes of soybeans and corn so they would resist the company's Roundup herbicide, whose active ingredient is glyphosate. As a result, Monsanto sold more Roundup as well as vast quantities of its herbicide-resistant seeds; by 2005, glyphosate-resistant soybeans filled 87 percent of US soybean fields.[2] Yet many consumers, concerned that such genetically modified organisms (GMOs) contain genes transferred from one plant (or bacteria) species to another, criticized GMOs as Frankenfoods. Perhaps more worrisome was that overuse of herbicides on GMO crops led to resistant superweeds and that seeds became increasingly controlled by large biotech corporations.

A less controversial result of genetic engineering is *Bacillus thuringiensis* (Bt). Scientists in the late 1950s discovered that this soil-based bacterium produces proteins, called Cry, that kill moths and other crop-destroying insects but cause little or no harm to beneficial insects, wildlife, or humans. Organic farmers for years sprayed this natural bug killer before large seed companies in the mid-1990s inserted the bacterium's gene (and its Bt toxic enzyme) into the genetic material of plants in order to make them resistant to certain pests. The resulting Bt crops essentially generate their own insecticides. US cotton farmers plant genetically modified Bt cotton on about three-quarters of their acreage to control bollworms, and corn growers deploy a slightly larger percentage to beat back the European corn borer and corn rootworm; as a result, synthetic insecticide spraying of corn over the past fifteen years fell tenfold, much to the environment's benefit.[3]

Early in the twenty-first century, Jennifer Doudna, a professor at the University of California, Berkeley, and Emmanuelle Charpentier, a French scientist, developed CRISPR-mediated genome editing that promises to cure genetic diseases; in 2020, they won the Nobel Prize in Chemistry for this discovery. Rather than modify or engineer genes from one species to another, the biochemists used a molecular-scissor-like process to edit genes in ways that make vegetables more nutritious or that develop treatments for diseases such as sickle cell anemia and cystic fibrosis. Doudna and Charpentier's innovation—the acronym CRISPR standing for "clustered regularly interspaced short palindromic repeats"—speeds up plant breeding but avoids the controversies associated with genetically modifying organisms. Yet some concerns remain; the International Summit on Human Gene Editing committee in 2016 declared that the technology should not be used to modify human embryos, and David Baltimore, president emeritus of the California Institute of Technology, warns that "scientists must be cautious, because once you have made a genomic alternation, there is no taking it back."[4]

CRISPR allows Doudna and other researchers to edit the DNA of a single species to advance or minimize a trait that already exists naturally. Admitting her comparison is a bit simplistic, the biochemist says CRISPR is like a computer's word processor that cuts and pastes text within a sentence. The major differences between gene editing and conventional breeding is CRISPR's swiftness, cost-effectiveness, and precision.

Bacteria edit their own genes, but scientists identified the natural process only in the late 1980s and began to unravel how it works in 2005. After investigating enzymes from more than seventy families of proteins, Doudna discovered one, known as Cas9, that works robustly within the CRISPR system to cut and paste RNA in desired ways. She and Charpentier in 2012 wrote the first academic paper on how to edit an organism's genes; their findings have been described as "one of the most monumental discoveries in biology,"[5] and *Time* magazine named Doudna one of the world's one hundred most influential people.

A less controversial alternative to transgenic engineering, this gene editing, known officially as CRISPR-Cas9, allows scientists to remove or add segments of RNA to repress or highlight certain traits. The cell then repairs the cut gene and passes on the preferred sequence to the next generation. Since CRISPR does not combine genes from multiple species, the US Department of Agriculture (USDA) in 2016 cleared for commercial sale a white button mushroom from which CRISPR-wielding scientists had extracted the browning gene, allowing the fungus to look whiter and last longer; not long after, the USDA signed off on a disease-resistant corn.

Gene editing is far faster than conventional breeding, irradiation, or genetic engineering. Commercializing a genetically modified organism (GMO) with a desired attribute, such as increased insect resistance, has taken an average of thirteen years and cost $130 million. Rapid scientific advances with CRISPR, computational biology, and artificial intelligence cut those numbers in half, and they keep falling with new technical improvements.

Still, the CRISPR process requires sophistication. A scientist must first map a plant's genome to identify and sequence the genes that prompt the desired attribute, be it increased tastiness or pest resistance. She then grows a few cells, edits them, grows the revised cells, turns those cells into full plants, and then makes the plants big enough for greenhouse and eventually field trials.

Rachel Haurwitz joined Doudna's Berkeley lab in 2008. A Texas native, she considers herself "an average science nerd"; as a teenager, she

Rachel Haurwitz, president and CEO of Caribou Biosciences. *Credit: Caribou Biosciences.*

tried to teach four hundred flatworms to navigate a maze in her parents' dining room. Haurwitz graduated from Harvard University, pursued her doctorate in molecular and cell biology at Berkeley, and at the age of twenty-five was tapped by Doudna to lead Caribou Biosciences, a new firm that licenses CRISPR technology to partners and that commercializes CRISPR-generated products. "One day was pipetting clear liquids from one tube to the other," the young scientist says, "and the next was trying to figure out how to build and grow a company."[6] Although Haurwitz still loves the discipline of biology, she maintains that she always wanted to be an entrepreneur who markets her scientific advances.

Doudna remains a cofounder and a member of Caribou's scientific advisory board. The professor argues that her former student is the perfect company leader because of her "deep knowledge of the opportunities for gene editing to solve problems in agriculture and health care."[7] Reflecting on the male-dominated biosciences field, Haurwitz asserts, "I like to believe that I and many others are proof positive that twentysomething-year-old women have a role to play in this industry."[8]

Although Caribou focuses mainly on treating cancer in humans, Haurwitz says that farms, with their bio-based products, will be a key economic segment changed by gene editing.[9] When the start-up turned to agriculture, it partnered with DuPont Pioneer, a giant seed corporation, to CRISPR-edit a waxy corn so it resists infection and tolerates drought. Caribou also joined forces with Genus, a United Kingdom livestock corporation, to raise pigs resistant to an aggressive disease—called porcine reproductive and respiratory syndrome (PRRS)—that prompts sows to miscarry and prevents piglets from breathing; PRRS costs European pig farmers almost $1.6 billion annually.

Located about a mile from the University of California, Berkeley, campus, Caribou Biosciences lies on a street with other single-level start-ups. Like many Bay Area entrepreneurs, the office features an open floor plan, complete with a Ping-Pong table, and a shiny, equipment-packed laboratory in the rear.

Haurwitz believes gene editing will transform agriculture. It is helping agronomists create plants that capture carbon, boost flavor, resist diseases, add nutrients, remove allergens, enhance affordability, and increase yields. CRISPR already produces reduced-gluten wheat that sensitive consumers can tolerate. It creates soybeans with no bad fat in order to provide a healthier oil for cooking, frying, and baking. It bolsters cacao's ability to avoid a virus that threatens the chocolate crop in West Africa.[10]

Sensitive to the stigma associated with GMOs, Haurwitz goes out of her way to explain that CRISPR does not create "transgenic" organisms. "Gene editing is not at all about taking DNA from a foreign species and integrating it into a plant," she repeats regularly. "It's really about working within the constraints of the plant's own genome."[11]

Although CRISPR may be the most popular gene-editing tool, the University of Minnesota created a separate approach, called TALEN (transcription activator-like effector nuclease), and licensed it to Calyxt, a biotech start-up headquartered a few miles northeast of Minneapolis. The company's name derives from the Latin word "calyx," for the sepals that surround, protect, and support a plant bud and flower.

Calyxt initially focused on soybeans and used TALEN to edit the DNA within two genes involved in the crop's fatty-acid synthesis. The resulting oil contains no trans fats and can be used for frying, cooking, and baking. To reduce farmers' risks, Calyxt provides the gene-edited seeds at no cost and then, at harvest, buys back the soybeans, which it crushes to extract high-value high-oleic oil; it feeds the remaining mush to pigs and cows.

Noting the TALEN-based product's longer shelf life and superior taste, Archer Daniels Midland in late 2020 agreed to buy all of Calyxt's high-oleic soybeans. The multinational food company crushes the beans and markets the resulting oil and meal. One independent study found that the gene-edited soybean oil enjoys three times the frying life of conventional soybean oils. Industrialists also use the bio-based liquid as a lubricant, high-performance motor oil, and solvent.

Calyxt's pipeline contains several gene-edited products, including a high-fiber wheat. The company also edits out much of the lignin in alfalfa, making the feed easier for farm animals to digest, and it sells potatoes that last longer in cold storage, thereby reducing food waste.

The start-up appreciates the sensitivities associated with manipulating genes. "In the U.S. there is widespread public distrust of GMOs," a former CEO says. "In part this is due to a perception that these products are unnatural and developed for the benefit of large corporations, not consumers." Yet he finds a growing number of consumers—particularly those with "health-related fears, such as diabetes, obesity, and food allergies that directly impact their food choices"—open to gene-edited meals.[12] Supermarkets already sell gene-edited soybeans and potatoes, and the USDA in 2018 declared that gene-edited crops are "indistinguishable" from those produced by traditional breeding and "do not require regulatory oversight."[13]

Yet not everyone is convinced. "This is the new kind of genetic engineering, whether you call it transgenic [GMO] or not," says the Center for Food Safety, an advocacy group. "It should be adequately regulated. We're not saying it should be stopped—we should know what has been done."[14] Others fear this more precise breeding method could produce unintended effects, and the European Union, long critical of genetically modified organisms, decided to regulate gene-edited plants.

Other skeptics argue gene editing enables Big Ag to become bigger and control food patents. No doubt giant food firms such as Coca-Cola and Archer Daniels Midland spend millions on sophisticated genetic editing, what they call synthetic biology, to create unique flavors and sweeteners.

Yet with proper training and basic laboratory equipment, gene editing is relatively simple and accessible. CRISPR "has been democratized," says Rodolphe Barrangou of North Carolina State University. "With 100,000 labs and 10 people per lab, we now may have over a million geneticists working with this technology."[15] Whereas GMOs force farmers and ranchers to buy expensive seeds from multinational monopolies, CRISPR enables growers in both developed and developing nations to enjoy more options in what they plant and produce.

A less serious concern, but still a legitimate one, is that major food processors will edit genes to provide only slight variations on the dreary taste of today's commercially produced fruits and vegetables, such as the bland tomatoes designed to fit exactly on a McDonald's bun. Yet others see hope: "There is something intriguing about using new technology to preserve the ravishing, sweet acidic burst of an heirloom tomato in a hardier, disease-resistant plant—an heirloom-plus, if you will."[16]

Leaving taste aside, CRISPR provides substantial environmental benefits. Snipping a plant's genome to favor disease-resistant genes cuts pesticide use, while other edits enable roots to absorb and hold carbon, thereby reducing the impacts of climate change. Gene editing also empowers efficient urban agriculture, the type done by Irving Fain's Bowery Farming. Said one botanist, we can "adapt the plant so that it's more compact, flowers faster, gives you a nice-sized fruit with a decent yield, in a very compressed growth setting, with the equivalent of protective agriculture—greenhouse conditions—but with LED lights."[17]

Creativity abounds in this emerging field that expands dramatically the age-old efforts to modify plants and animals. Gene editors create seeds resistant to droughts or floods associated with global warming. They develop tomatoes without seeds and other varieties resistant to powdery mildew. They increase the size of sweet potatoes by enhancing the flow of sugar to their roots. They remove caffeine from coffee, avoiding today's costly and taste-reducing decaffeination process.

Innovative biology, Doudna declares, is "powerful as a tool for manipulating the genetic material in plants. And so, we have opportunities now in agriculture that were never available previously."[18] Said one grower who marvels at what the professor's discovery means for farms and foods, "If we are going to make a healthier U.S., it has to start with us."[19]

CHAPTER 22

Lee DeHaan, The Land Institute— Planting Perennials

Agronomists in the nation's agricultural heartland for decades have sought to develop a grain that needs neither tilling, synthetic fertilizer, nor pesticide. More recently, they added the goal of having that plant sequester carbon to mitigate global warming.

Wes Jackson and his colleagues at The Land Institute, a nonprofit based in central Kansas with a 900-acre research campus, focus on perennial crops that live more than two years and whose deep roots capture and store carbon dioxide. Their key product is Kernza, a hybrid intermediate wheatgrass with the scientific name *Thinopyrum intermedium* that offers sweet and nutty kernels attractive to bakers and brewers.

A leader of the sustainable-agriculture movement, Jackson is a farmer, geneticist, philosopher, author, and recipient of the Right Livelihood Award and MacArthur Fellowship. Born and raised on a farm near Topeka, Kansas, he earned his doctorate in genetics and established an early university-based environmental studies program. He left academia in 1976 to start The Land Institute and develop perennial cropping that cuts tillage, reduces topsoil loss due to erosion, slashes chemical runoff, and sequesters greenhouse gases; he calls his work "an inversion of industrial agriculture."[1] Although retired from The Land Institute's presidency, he remains involved in its wheat, sorghum, and sunflower initiatives.

Now leading the organization's Kernza project is Lee DeHaan, who grew up on a corn and soybean farm in southern Minnesota but became skeptical of the consistent plowing and heavy spraying of pesticides and fertilizers. He obtained his doctorate in agronomy and applied plant science from the University of Minnesota, and he now oversees a $3 million per year plant-breeding program that tries to improve the hardiness of perennial wheatgrass.

The lanky scientist displays an earnest passion for crop breeding. "When you select plants to mate generation after generation, you begin to feel that these families of plants are in some strange way your adopted children," he says. "It is really exciting to see them out in the world on their own living healthy and productive lives!"[2]

Kernza has become a key family member for DeHaan. As a perennial, the crop gets harvested several times before dying. In contrast, farmers plant almost 70 percent of global croplands with annuals—including cereals, oilseeds, and beans—that die after one growing season. Since Kernza does not need to be plowed under each year, its use cuts erosion of precious topsoil by up to 50 percent. Compared with industrialized agriculture's annuals, it reduces nitrogen runoff from chemical fertilizers thirty-five-fold. Kernza consumes five times less water, reducing the strain on groundwater supplies. It demands fewer fossil fuels burned in tractors and cultivators; one study found that widespread farming of perennial corn would cut US diesel costs by $300 million annually.[3] Kernza emerges early in the spring and overwhelms weeds, thus dropping the need for herbicides. Its ten-foot-deep roots, like those of the prairie grasses that once covered the Midwest, also sequester atmospheric carbon dioxide.

Despite these advantages, Kernza kernels measure about one-fifth the

Lee DeHaan, lead scientist for The Land Institute. *Credit: The Land Institute.*

size of conventional spring wheat, and the plant does not grow well in warm southern latitudes. Yields, moreover, are relatively small, so the grain's current uses are limited to niche markets. DeHaan's breeding goal, therefore, is "to encourage the plant to reallocate to crop development the energy and resources that would have gone to growing stems and roots."[4]

Because grain yields decline after a few harvests, DeHaan suggests that farmers use the intermediate wheatgrass for several additional years as hay for their livestock. This dual use, says the scientist, "can push perennials to be profitable" as well as "extend the amount of time the land is not disturbed."[5]

Breeding perennials has been unsuccessful before. Soviet researchers in the 1930s tried to hybridize perennial wheat but abandoned the effort with little progress three decades later. California scientists took up the effort and produced promising perennial prototypes, until the green revolution increased outputs from annual crops.

Since 2003, The Land Institute has been trying to develop within Kernza certain attributes—better yields, larger seeds, and synchronous flowering and seed maturation. The group's breeders select and cross-pollinate plants exhibiting those characteristics, they plant the resulting seeds for the next generation, and they repeat the process until the crop demonstrates the desired attribute. Such breeding is a slow but time-tested process.[6] Sparking optimism, however, are new technologies, such as genomic selection (a rapid test that predicts a plant's characteristics and yield) and CRISPR (a gene-editing process highlighted in chapter 21). DeHaan refers to these as "excellent and powerful tools that accelerate and focus traditional breeding methods."[7]

Other off-season crops, such as *Camelina*, flax, and Kura clover, can produce some of the same benefits. Don Wyse, a University of Minnesota professor, complains that conventional corn and soybeans use farmlands for only about three months every year, and he argues that cultivators need other opportunities to make a profit. Year-round croppings, he adds, reduce erosion and capture carbon.[8]

Jerry Steiner, CEO of CoverCress in St. Louis, Missouri, converts native pennycress, a flowering plant in the cabbage family, into an off-season crop rich in proteins. He sows the seedlings in the fall, and the pennycress grows until the spring's regular planting of corn and soybeans. Over the winter, Steiner's new crop reduces erosion and retards weeds, and when harvested, it provides both oil and meal for livestock—and additional revenue for farmers. Claiming to be a conservationist at heart, this innovator boasts that off-season cover crops offer environmental benefits and solid economics.[9]

DeHaan focuses on Kernza grain, but other Land Institute researchers work on perennial rice, sorghum, and sunflowers. In addition to conducting scientific investigations, they network with chefs to concoct creative recipes that demonstrate perennials lead to delicious meals. Food processors tend to blend Kernza with annual wheat flour to bake bread. Cooks use the intermediate wheatgrass by itself to make muffins, pancakes, and a pilaf-like dish.

Yvon Chouinard, the founder of organic clothing maker Patagonia, helps market Kernza products. Applauding the crop's potential to cut CO_2 emissions, Chouinard declares Kernza to be "the most important experiment we've ever tried."[10] Expanding on his purchases of sustainable hemp, cotton, and wool for clothing, Patagonia's organic-food subsidiary buys the wheatgrass grain to produce Long Root Pale Ale, which it brews in Portland, Oregon, and sells in pint-size cans at Whole Foods Market stores on the West Coast.

Also promoting DeHaan's perennial is General Mills, which markets Cascadian Farm Honey Toasted Kernza cereal. Having pledged to cut its greenhouse-gas emissions by 28 percent within a decade, the food giant estimates that expanding this cereal to be 1 percent of its grain-based products would create a multimillion-dollar market.

General Mills also made a charitable contribution that allows The Land Institute and the University of Minnesota to measure greenhouse-gas reductions from perennial plants, as well as to investigate how the quality and quantity of perennials vary with fluctuating weather, including winter chills, spring freezes, and summer heat stresses. DeHaan welcomes the contribution and cooperation but hopes private-sector investments soon will support such agroecological research.

DeHaan recognizes that to meet the demand from the Patagonia and General Mills purchases, production of this intermediate wheatgrass must expand dramatically. Seed companies are logical marketers to farmers, but the giant grain-seed corporations tend to fear that Kernza will disrupt their businesses. Smaller firms, however, are very interested and see an opportunity to expand their offerings beyond perennial grasses to grains. While claiming "there's nothing with greater promise out there," DeHaan admits Kernza is not the silver bullet that alone will reverse climate change and feed the world. He receives numerous calls from interested farmers, but he acknowledges, "I spend much of my time trying to keep people from becoming too excited."[11]

Much work remains. Most obviously, DeHaan and other growers need to develop more perennials and market more foods that entice more consumers. On the research side, DeHaan wants to understand Kernza's sequestration capacity over an extended period and its ability to stave off diseases. The scientist, for instance, has shown that the intermediate wheatgrass repels leaf or stem rust, among the most common wheat diseases, but he hopes to develop new strains that also resist shattering, when high wind or rain strips mature seeds from the plant.[12]

Although Kernza has only entered niche markets and attracted limited private-sector investment, the optimistic DeHaan argues, "Perennial grains, legumes and oilseed varieties represent a paradigm shift in modern agriculture and hold great potential for truly sustainable production systems."[13] Perennials, he continues, "simultaneously address farming's three challenges—solving environmental problems, providing food for people, and creating reliable incomes for farmers." While recognizing "time frame and technological obstacles," DeHaan declares: "The payoff is so big that the effort is worthwhile."[14]

CHAPTER 23

Joshua Goldman, Australis
Aquaculture—Blocking Burps

Cattle belch a tremendous amount of climate-changing gas. When the animal chews grass, hay, cornstalks, or virtually any food, microbes living in its rumen, the first and largest chamber of its four-part stomach, ferment that forage and provide the ruminant with nutrients. Yet this microbe-feasting digestion also releases methane, a potent pollutant that the animal burps quite frequently into the air.

Belching cattle, by some estimates, release more greenhouse gases than leaking oil and gas wells, coal mines, festering bogs, or burning forests.[1] A few numbers explain the enormous impact of such enteric fermentation. Each cow or steer burps about four hundred pounds of methane every day, equal to the climate-change damage from an automobile driving across the United States four times. Methane, moreover, is potent; despite having a shorter life span than carbon dioxide, it is eighty-four times more damaging in its first twenty years. The world's cattle number almost one billion, which means livestock belching causes between 14.5 percent and 51 percent of global greenhouse-gas emissions. (Cattle farts have a similar, but smaller, consequence. Other methane-releasing ruminants include sheep, goats, and buffalo.)

Although it sounds strange to say, tackling climate change requires less animal burping. "We domesticated ruminants over 10,000 years ago and relatively little has changed," explains one scientist. "It's time that got an upgrade."[2] Ranchers for decades tried numerous cattle-feed additives, including oregano and tea leaves, but methane reductions proved to be modest. Yet in the past few years, innovators, including Joshua Goldman, have used science to devise fresh, and far more effective, belch blockers.

Goldman for thirty years championed climate-smart fishing. That mission began in his Hampshire College dorm room, where he located a small pond heated by a nearby greenhouse. At the age of twenty, Goldman secured a $400,000 grant from The Pew Charitable Trusts for his aquaculture research. Upon graduation, he cofounded Bioshelters, whose aquaponics farms nurtured tilapia, a fish that became increasingly popular as a source of proteins and nutrients.

The entrepreneur, however, was not satisfied and spent three years searching for a better fish to farm, investigating more than thirty species before settling on barramundi. This Asian sea bass, roughly translated from an Aboriginal language as "large-scale river fish," is widely distributed throughout the seas around South Asia, Papua New Guinea, and Northern Australia.

From a health perspective, Goldman favors barramundi because it contains half the calories of salmon but far more proteins and omega-3 fatty acids, which promote cardiovascular and brain health. From an environmental standpoint, farming barramundi requires six times less fish-based feed than salmon, and it does not cause the oxygen-poor dead zones that result when salmon feces and other wastes fall to the bottom of their large pens in the open ocean. Barramundi's white flesh resembles that of snapper or sole, tastes "sweet and buttery with a delicate texture," and is sold throughout North America, mostly in Asian markets.[3]

Joshua Goldman, CEO of Australis Aquaculture. *Credit: Australis Aquaculture.*

Goldman in 2004 formed Australis Aquaculture, which initially harvested about 3,500 barramundi each day from tanks the size of swimming pools within former airplane hangars in Turners Falls, a village in Western Massachusetts's Pioneer Valley. Using Goldman's multiple patents, the firm filled those tanks with water from the nearby Connecticut River and recycled it three hundred times, separating and sending the fish's solid wastes to local farmers as fertilizer. The entrepreneur received awards for innovation and sustainability from the Institute of Food Technology.

In order "to sustain ourselves and restore our planet by unlocking the potential of sustainable ocean farming," Australis Aquaculture in

2006 began farming barramundi in pens off the coast of Vietnam.[4] It expanded that operation in 2016 with the acquisition of Marine Farms Vietnam and became that country's largest aquaculture sea-lease holder.

Despite a successful barramundi operation, Goldman again sought a new challenge. As an aquaponics farmer, he long had cultivated seaweed in his fish farms because the macroalgae absorb nutrient wastes, provide habitat for small marine life, and create good harvesting jobs in the adjacent rural areas. He initially raised *Kappaphycus*, a seaweed genus that provides an extract used to thicken many food products.

Yet Goldman began growing numerous other species, ever "on the lookout for other types of seaweed that might more fully allow us to realize our vision for regenerative ocean farming."[5] He had heard about the environmental damage from livestock burps and the potential for "green grazing." Yet what spurred him to establish a new initiative within Australis was a 2014 study by James Cook University and the Commonwealth Scientific and Industrial Research Organisation (CSIRO, Australia's scientific-research agency) that found a ruminant's greenhouse-gas emissions fell dramatically when 1 percent of its diet included red seaweed. The researchers tested twenty different types of red seaweed and found that one species, *Asparagopsis taxiformis*, slashed such emissions by up to 99 percent.[6] Although skeptics doubt such a huge reduction, scientists at the University of California, Davis, demonstrated 80–90 percent emission reductions at 0.2–0.5 percent feed-inclusion rates. According to Goldman, "This unprecedented level of reduction is due to *Asparagopsis*'s unique and highly complex chemistry—involving over one hundred naturally occurring bioactive compounds that appear to work synergistically"[7] and that interact with a vitamin B_{12} cofactor in the cow's rumen.[8] Since that red seaweed thrives in tropical and subtropical waters, Goldman felt its cultivation would complement his Vietnam-based barramundi farming.

The entrepreneur claims to be launching an "aquatic moonshot" that combats climate change by "transforming beef and dairy into sustainable industries."[9] Blocking methane burps from the world's one billion cattle would offer the equivalent of blocking emissions from every vehicle on the planet. Although such a goal seems massive, Goldman tries to suggest its simplicity by saying, "The cool thing is that farmers only need less than one percent of the seaweed in their livestock feed to significantly reduce a cow's methane emissions. We're not out there trying to replace grass or corn—we're just sprinkling a little bit of salt and pepper on livestock feed to help the planet."[10]

Goldman targets the red seaweed for ruminants, not humans. When *Homo sapiens* eats seaweed, we tend to favor kelp, a variety increasingly farmed in the cold waters off the coasts of Maine and Japan and used in salads and mixed into a variety of common foods and household items. Kelp requires no arable land, fresh water, fertilizers, or pesticides; in fact, seaweed farming improves water quality and sequesters carbon dioxide, phosphorus, and nitrogen. A growing number of chefs praise kelp's tastiness and its rich combination of iron, calcium, fiber, and vitamins.

Goldman recognizes the challenges of transitioning from belch-cutting studies to a viable business. He needs, for instance, to collect and characterize diverse substrains of *Asparagopsis* and create multiple methodologies for ocean-based cultivation. Although Goldman has mastered how to entice seaweed spores to stick to ropes that form the basis of underwater farms, he must vastly increase the scale of his production. He also requires more research on two potential problems. First, some cows don't seem to like eating seaweed, and their milk production drops; second, one study found worrisome increases in iodine and bromine in the milk of seaweed-eating cows.

Australis is looking for more places to grow the red seaweed. Goldman calculates that *Asparagopsis* farms covering 4 percent of the world's oceans would cut greenhouse gases almost in half, but that is an enormous expanse. There are two species of this pinkish-red sea plant, one that grows in temperate waters and the other in more tropical waters. Goldman says, "These are native and could likely be farmed in over seventy countries globally."[11]

The innovator also must convince feed processors and cattle owners to try belch-reducing additives. Since the meat industry is not vertically integrated, meaning large packers do not own the cattle, the entrepreneur's marketing must be widespread. Yet Goldman maintains, "Integrating seaweed culture with ocean-based fish farming to improve the diets of ruminant animals may be one of the most effective systems to combat climate change."[12] He also maintains that *Asparagopsis* can make ranching much more profitable, since it reduces feed requirements, allows producers to sell carbon offsets, and develops a more environmentally benign product for which consumers and brands will pay a premium.[13]

Another effort to make livestock less gassy comes from DSM, which began in 1902 as Dutch State Mines and now is a 21,000-employee diversified biotech giant. After its ten-year-long, research-focused Project Clean Cow, DSM scientists settled on a feed additive that hampers the ruminants' methane-producing enzyme. Marketed as Bovaer, that supplement is known to scientists as 3-nitrooxypropanol, or 3-NOP. Independent studies find it cuts methane emissions by 23–28 percent.[14]

Mootral, a Swiss start-up, offers one other belch-reducing enhancement. Its German-born CEO and cofounder, Thomas Hafner, appears to be a classic entrepreneur; he dropped out of college, worked for a short time at a Burger King restaurant, and then created a life sciences company that marketed over-the-counter diet, antacid, and allergy supplements. He sold that business in 2014 for $150 million and decided to do for a cow's digestive tract what he had done for people.

Being familiar with garlic's antimicrobial properties, Hafner began testing numerous combinations of its extracts. He claims that Mootral's proprietary mix of garlic and citrus, called Ruminant, cuts methane emissions from cattle by 58 percent in laboratory experiments and up to 38 percent in field tests. Although independent researchers in Denmark and Germany reached similar conclusions, one study found that methane emissions from extract-eating cows were "not significantly different."[15] Yet even these mixed results attracted venture capitalists, including Chris Sacca, who made billions financing Twitter and Uber, and Hafner himself invested $20 million of his own money.

Methane reductions result from the feed supplement inhibiting the activity of archaea, a specific group of gas-producing microbes in the cattle's rumen, without harming the bacteria that digest the fibrous food. Hafner compares cows ingesting the garlic-and-citrus extract to humans gulping Milk of Magnesia to neutralize a gassy stomach.

To test this natural dietary supplement, Mootral initially focused on four hundred cows located at a dairy farm in a former coal-mining region of Wales. Researchers first had to determine the correct dosage; too much of the garlic-citrus mix, for instance, hurts a cow's ability to digest food and makes its milk taste garlicky. They calculated that the pellets increased dairy yields by 8 percent, probably because cows that spend less energy burping have more energy to produce milk.[16] Scientists also discovered that the mixture enhanced the liquid's nutritional quality with more fatty acids and proteins, reduced udder infections (and therefore cut costs for antibiotics), and swatted down the number of flies that bothered the animals (possibly because the insects dislike the cows' garlic breath). The company's investigators hope soon to release methane-reducing mixtures for other ruminants, including sheep and goats.

Mootral's Ruminant comes in powdered form and can be added to feed pellets at the mill or mixed on the farm. The company recommends 10–15 grams per day per cow, depending upon the breed and weight, a tiny portion of a cow's daily food intake of 75–110 pounds. The supplement costs about $60 per cow per year (or 16 cents per day), which the company claims is offset by increased milk output. CEO Hafner also argues that eco-conscious customers and supermarkets will pay a bit more for the liquid's climate benefits. A clever British dairy, Brades Farm, markets its milk from Mootral-supplemented cows as "low methane" and "Less CO_2W Burps."

Mootral realized another revenue source in late 2019 when it obtained greenhouse-gas credits in the carbon-offset market. Verra, which runs the largest exchange for voluntary emission reductions, agreed that ranchers and dairy farmers using the garlic-and-citrus feed supplement deserve compensation for lowering methane gases burped out by cattle. In the past, most emission credits went to foresters planting trees that absorb carbon dioxide or wind generators that displace coal-fired power plants. Although the voluntary carbon market in 2018 totaled only $300 million, it may grow as more countries mandate greenhouse-gas reductions to meet international climate agreements.

Hafner acknowledges that he needs more peer-reviewed studies to convince ranchers, dairy farmers, and consumers of the benefits of "climate-smart cows" under varying farming conditions, cattle breeds, and livestock diets. To tap carbon markets and claim "cow credits," he also must continue to pass strict auditing processes that ensure environmental credibility. The company, moreover, must vastly increase production and keep costs down. Still, the CEO argues that Mootral shows that sustainability and profitability can go hand in hand—for all stakeholders involved.[17]

The combination of Mootral, DSM, and Australis demonstrates the enormous inventiveness and effort in the past few years to reduce methane emissions from animal agriculture. Seeking simple solutions to a massive problem, these quite different innovators reflect the rapid transition from research to market-based ventures. Speaking about the benefits of feeding methane-reducing supplements to cattle, Josh Goldman concludes, "You don't have to rebuild 10,000 power plants in the world [to combat climate change]. You basically create a modest feed additive that has a big leverage effect."[18]

CHAPTER 24

Julia Collins, Planet FWD—Creating a Climate-Friendly Food Platform

Julia Collins calls herself a founder, the Silicon Valley term for a serial entrepreneur. She knew early on that she would focus her businesses on food. Her grandfather operated a little hothouse that grew tomatoes in the Noe Valley neighborhood of San Francisco; her grandmother always seemed to have something on her stove, including her cheese grits specialty; Collins created her own hydroponic garden in the sixth grade; and the following year, she founded a composting club called WORM (World Organic Recycling Movement).

After graduating from Harvard University, with a degree in biomedical engineering, and Stanford Graduate School of Business, she briefly ran a New York City restaurant before cofounding Zume Pizza in 2015. She considered the start-up to be a high-tech food-delivery innovator.

More and more foods get delivered. Pizza long has been a dorm-room tradition, but now DoorDash and Uber Eats bring a multitude of restaurant meals to our homes, while Safeway and Amazon convey supermarket produce and meats. UPS predicts that over the next decade, the food-delivery business will grow more than tenfold, from $30 billion to $360 billion.

Finding such food deliveries to be little more than taxi services, Collins identified two meal-development problems—food demand varies by time of day and area, and no traditional brick-and-mortar locations offer an ideal mix of breakfast, lunch, and dinner traffic. Her solution: mobile kitchens that move close to probable consumers on the basis of predictive analysis. High-tech food vans, she envisioned, would cut the time between cooking and eating, reduce food waste, and offer an alternative to traditional restaurants.

Brick-and-mortar eateries usually take two years to locate, permit, and build. Zume created an alternative that gets cooks to markets faster and with less cost. The start-up also allowed penetration into microtrade areas—such as office parks, hospitals, and stadiums—that offer high consumer traffic during specific periods but low volumes overall.

Zume saw itself as a robotics company that happened to deliver pizza. It used predictive algorithms to anticipate who would likely order which pizza and when, and then it strategically located a fleet of mobile kitchens, essentially thirty-foot-long GPS-equipped food vans with fifty-six automated ovens, to be within a half-mile radius of those consumer concentrations. When an order came in via app, computer, or phone, Zume robots quickly moved the correctly topped pizza from a preheater to an oven, which cooked the pie en route and timed it to be ready upon arrival at the customer's address. A self-cleaning robot finished the job and sliced the pie into the requested number of pieces. The mobile kitchen, always close to the customer, delivered a piping hot pizza within minutes of its order.

A transition from home cooking to food delivery, said Collins, suggests numerous consequences beyond warmer meals. With fewer people cooking, we may need to shrink and redesign kitchens. Fast-moving delivery services could challenge traditional supermarkets and conventional diners. Yet since many meals may not travel well, the delivery boom could further limit the diversity of food choices.

Julia Collins, CEO of Planet FWD. *Credit: Planet FWD.*

Building on this pizza-delivery experience, Zume in 2018 began to license its predictive analytics and automation technology to help other brands make their food-supply chains smarter. Providing "actionable" information, the start-up enabled marketers "to expand operations closer to customer demand and use resources more efficiently." Zume also helped food preparers grow their revenue with multiple menus and mobile-kitchen configurations for breakfast, lunch, afternoon snacks, dinner, and evening treats.

Collins and Alex Garden, Zume's cofounder, attracted high-profile investors, including Jerry Yang, cofounder of Yahoo!; the Japanese conglomerate SoftBank Corp. provided $375 million in November 2018;

and analysts in early 2020 estimated the company's value to be \$2–\$4 billion.[1] Headquartered in Mountain View, Zume had offices in San Francisco and Seattle and planned for expansion.

Garden, unlike his partner, did not grow up thinking about food or farms. He got his start developing video games and digital media, and for several years he managed Xbox at Microsoft. More relevant to this food-delivery venture, Garden founded 1VALET, which provides amenities to multifamily housing. When not launching start-ups, he builds race cars and motorcycles in his home machine shop.

Garden increasingly focused on packaging that would ensure delivered food arrives at the right temperature and with quality presentation. To limit its environmental impact, says Zume's cofounder, cartons should be made of recycled material, be recyclable, and not contain petrochemicals. Zume's patented offerings, which cost less than plastic bags, are sturdy boxes made of molded fiber constructed from the stalks of sugarcane and other agricultural discards. Zume also provides Pizza Hut with round compostable boxes that use less packaging because they lack corners; the firm claims its boxes also interlock easily "to ensure a smoother delivery."

Collins in November 2018 decided to leave Zume, which she claims was a tough decision. She doesn't discuss her move or what happened with the firm, which in mid-2020 laid off 80 percent of its staff and shut down its pizza-making and delivery business and turned solely to food packaging.

Collins took several months to consider what to do next. She says the combination of becoming a mother, turning forty, and feeling momentum from being a Silicon Valley founder caused her to think big. "When I finally got the conviction to follow my passion, I just didn't let anything stand in the way," says Collins. "I knew that it was food and anything that wasn't food was a distraction."[2]

That focus led Collins to climate change, since industrial agriculture releases so many greenhouse gases. She founded Planet FWD to cut that pollution, help farms sequester carbon dioxide from the atmosphere, and make it easier to bring climate-friendly food products to the market. She is not modest about her ambitions: "I'm out to protect the planet and feed the world."[3]

With a $2.7 million seed round led by BBG Ventures, Collins began to create a sustainable-food platform, an extensive archive about farms engaged in regenerative agriculture. That database enables food processors to work with carbon-sequestering operations and produce climate-friendly foods. Planet FWD also develops its own products, the first being crackers and chips, that have low carbon footprints.

"As I started to pull together the ingredients for my climate-friendly snacks," says Collins, "I amassed this really exhaustive library of all this information about these [greenhouse-gas-sequestering] farms. And I thought that was really interesting because anybody who wants to create a climate-friendly food product needs a universal set of information that just wasn't available."[4]

Collins hopes increased consumer demand for climate-friendly products will motivate more farmers to sequester carbon through cover crops, deep-rooted perennials, compost and manure, livestock integration, plant diversity, and crop rotations. Focused on soil health, she also argues for farming practices that move away from the synthetic fertilizers and intensive plowing that stimulate nitrous-oxide emissions.

In addition to mitigating climate change, these approaches—what Collins dubs "beyond organic," "sustainable," or "regenerative"—return minerals to depleted soils, avoid erosion, keep pests off balance, and stimulate soil fertility. One study found that an increasing, but still relatively small, number of growers deploy these actions, and it suggested that we "may be approaching a tipping point where it could be transformative."[5] Collins argues that her software brings the high tech needed for preindustrial husbandry to expand.

Having little trust in politicians to move the needle on global warming, Collins believes in "the power of private companies to make a change." Her goal may be more ambitious than even the most bullish estimates, but she asks that we "imagine a world where all food companies shifted to a regenerative organic agriculture system. That would actually get us to a point where we were carbon negative. Where not only were we not increasing the amount of greenhouse gases, but we were actually sequestering it all back, just through agriculture."[6]

Collins has built a multibillion-dollar company and attracted investors to a new venture, but she admits, "Being a Black woman in Silicon [Valley] can really give you imposter syndrome. When you drive up to Sand Hill Road and open those doors and you do not see a single person who looks like you, it can really shake your confidence."[7] Statistics back up her concerns. In 2018, US venture capitalists invested $131 billion in nearly nine thousand start-up businesses, but only 14 percent had at least one female founder,[8] and less than 1 percent went to women-of-color CEOs.[9] By sharing her own difficult experiences and connecting young entrepreneurs to investors, Collins works "to make the world more equitable" by "accelerating the success of Black women, of intersectional people, of people of color in these spaces."[10]

Collins calls herself "a dreamer," an entrepreneur willing to walk away from a successful venture and build something new. Her vision for a database of sustainable ingredients, she says, is to "connect farmers who are helping to bury carbon in the soil with consumers who are hungry to take action on climate change." Collins says, "Planet FWD is on a mission to help undo climate change by making it easier to bring climate-friendly food products to market."[11]

CHAPTER 25

Stafford Sheehan and Gregory Constantine, Air Company— Cutting Carbon with Vodka

Cocktail hour is not usually associated with saving the planet. Yet Stafford Sheehan has figured out how to make a bottle of alcohol soak up as much carbon dioxide as eight fully grown trees. He and his business partner, Gregory Constantine, formed Air Company in 2017 and located their operations in an old warehouse within the Bushwick neighborhood of Brooklyn. They claim Air, this innovative spirit, is "the world's first carbon-negative vodka" and that their "carbon conversion technology could make a substantial difference in the fight to end climate change."[1]

With a doctorate from Yale University and applied research experience in chemistry, physics, and computer science, Sheehan patented a system that uses solar electricity to turn carbon from the air into alcohol. While conventional vodka production emits about thirteen pounds of greenhouse gases with every bottle, his product removes a pound.

Sheehan and Constantine met in 2016 at a bar in Israel, where the two were on a trip sponsored by *Forbes* magazine for entrepreneurs under thirty. They hit it off, and when they ran into each other again at a different bar, Sheehan displayed an unusual bottle of booze. According to Constantine, then an executive at Diageo—a spirit and beer producer whose brands include Smirnoff, Johnnie Walker, Baileys, CÎROC, and Guinness—"I said, 'Hold on, you made this from carbon dioxide?'"[2]

The technology to turn carbon dioxide into ethanol is not new. Several companies use it to make fuel and other products. In the agricultural space, Solar Foods of Finland creates a protein bar from the greenhouse gas, and Kiverdi, a start-up based in the San Francisco Bay Area, offers a sustainable alternative to palm oil.

Sheehan, however, claims Air Company has developed a "secret sauce" that expands production substantially and sustainably. The firm begins by using solar-powered electrolysis to split water into its components, hydrogen and oxygen. It then combines those gases with leftover carbon dioxide from nearby factories and places the mixture into a special reactor that emits oxygen and produces alcohol.

That approach, says Sheehan, is "inspired by photosynthesis in nature, where plants breathe in CO_2. They take up water, and they use energy in the form of sunlight to make things like sugars and to make other higher-value hydrocarbons, with oxygen as the sole by-product. Same thing with our process: The only by-product is oxygen."[3]

Air Company's vodka-production process took two years to perfect, including overcoming some serious regulatory and construction challenges. The company initially had to convince skeptical Brooklyn fire chiefs to permit an alcohol-making factory in their neighborhood of wooden warehouses and homes. Once they got over that hurdle, they arranged for a special crane to drop the multi-ton conversion reactor—Sheehan's "secret sauce"—through the warehouse roof.

Traditional distillers make vodka from fermented corn, rice, wheat, or potatoes. Boasting of their own product's environmental benefits, Sheehan and Constantine note that growing those crops requires thousands of acres of farmland, tons of synthetic fertilizers and pesticides, and vast releases of greenhouse gases.

Gregory Constantine (left) and Stafford Sheehan (right), cofounders of Air Company.
Credit: Air Company.

Air Company's motto is "Goods that do good." The young entrepreneurs claim their carbon-negative vodka is more pure than traditional spirits, since conventional distillation does not completely remove methanol and carbolic acid. According to Constantine, "Air Co.'s process circumvents the production of these impurities entirely, by connecting two carbon dioxide molecules together—'building up' to produce ethanol, rather than 'breaking down' larger molecules that produces a wash with high impurity content."[4]

With experience in the music and consumer goods industries, Constantine appears to be a natural marketer. He bills Air Company as a "modern luxury lifestyle brand"[5] and its product as the world's "most sustainable spirit."[6] He attracted dozens of reporters to the firm's launch, boasting in one interview, "I've always been so drawn to innovating in everything I do, and trying to inspire others through purpose and through self difference."[7] Constantine claims "design is one of the biggest pillars of Air Co.," and he carefully drafted the container's label to declare that this vodka is made from air, water, and sun, that the packaging is recyclable, and that the label's custom-made removable glue allows the 750-milliliter bottle to be reused.

Sheehan and Constantine think beyond vodka. Their ability to make pure ethanol from carbon dioxide could create a variety of products, including fragrances, fuels (particularly high-value rocket propellant), home-cleaning supplies, and rum and other spirits. During the 2020 coronavirus pandemic, for instance, Air Company produced and donated carbon-negative hand sanitizer, which is composed of 70 percent ethanol, Air Company's primary output.[8]

Air Company also works with the National Aeronautics and Space Administration to convert the carbon dioxide that astronauts exhale into glucose for food or fuel during long-distance flights. Such recycling helps the space agency close material loops on tight-quartered space stations and lends promise for future missions. The company helps NASA on the ground, too, working with the agency and the United Nations to produce clean fuels for cars, planes, and rockets. For its efforts, the firm won NASA's CO_2 Conversion Challenge and the United Nations' Ideas4Change Award.

Like many start-ups, Air Company worries about production capacity. In 2020, it sold vodka bottles to a few high-end restaurants and neighborhood bars not far from its Brooklyn headquarters. Yet it is building a manufacturing facility in Ontario, Canada—with ten times the capacity of its Brooklyn distillery—and scouting other locations around the globe. "The benefit of this technology is it is extremely modular," says Constantine. "What we're able to fit into a 500- to 1,000-square-foot space, traditional alcohol production methods and distilleries would need football fields and football fields of corn and irrigation. We can do that in a very metropolitan area, and that allows us to potentially displace transportation [costs and pollution] by placing these, hopefully, around the country."[9]

The company also frets about whether enough consumers will pay three times the price of popular brands for eco-friendly spirits. Several journalists say the carbon-negative liquid tastes like other vodkas, yet perhaps with a pleasant hint of ethanol. It won a gold medal in the Vodka—Ultra Premium round, with one taster declaring, "I loved the texture of this vodka, which had a little viscosity."[10]

Air Company will not discuss its costs, but the cofounders know that to attract more investors and customers, those outlays must fall as production rises. They claim to be building a technology company that turns carbon into value. "We are on a mission to change the world and save the planet," declares Constantine. "You have to start somewhere, and we are starting in vodka."[11]

Disrupting Farms and Foods

No doubt a few computerized farm systems have been around for a decade or two; Deere & Company, for instance, introduced GPS-guided tractors in 1997. Yet modern technologies help growers do far more than plant straighter rows. Today, entrepreneurs are adapting recent and rapid advances to redefine the very nature of food and farming.

Ag-tech innovators—those profiled here and many more—arrive at a time when malnutrition and environmental calamities grow; when green revolution developments of the mid-twentieth century—synthetic fertilizers, chemical pesticides, and genetically modified organisms (GMOs)—plateau; when Big Ag seems stuck in the status quo; and when preindustrial practices struggle to gain acceptance. Today's entrepreneurs think it profitable to remake agriculture.

They promise "creative destruction," a term first coined in 1942 by Austrian economist Joseph Schumpeter to describe how innovation improves society by replacing outmoded practices. The quintessential example is Henry Ford's assembly line, which destroyed the jobs and businesses tied to horse-drawn carriages. (Ford acknowledged his inspiration came from Chicago's meat-processing plants, those that Upton

Sinclair portrayed in *The Jungle*; author Jonathan Safran Foer observes, "Putting together a car is just taking apart a cow in reverse.")[1] For Schumpeter, creative destruction was as natural and inevitable as evolution. He famously wrote, "The process of industrial mutation . . . incessantly revolutionizes the economic structure from within, incessantly destroying the old one, incessantly creating a new one."[2]

While several of the innovators profiled in this book boast that they will fully supplant Big Ag, even skeptics of such bold claims acknowledge that emerging technologies challenge agriculture's oligopolies and open the food industry to needed competition.[3] Noting that huge agribusinesses (and large farm-focused bureaucracies and land-grant universities) arose in the previous century to provide stability rather than to respond to fast-moving challenges,[4] there are already signs that upstarts can shake conventional strangleholds on food markets. In addition to the profiled entrepreneurs, one notable example is the alternatives to Big Dairy's cow milk; sales of soy, almond, oat, rice, and other plant-based substitutes rose from being a niche market less than a decade ago to having sales of $22.6 billion in 2020 (and projected to reach $40.6 billion by 2026);[5] cow-milk consumption per capita, in contrast, fell by roughly 40 percent,[6] and twenty thousand dairy farms went out of business.[7] Such trends forced conventional giant Dean Foods to declare bankruptcy in November 2019, and more innovation and competition are on the horizon. Without raising cows, California-based Perfect Day makes dairy products—including milk, yogurt, ice cream, and cheese—by fermenting a natural type of microflora to create the proteins, specifically casein and whey, historically found in milk. Promoting its dairy protein as animal-free and lactose-free, the company raised an additional $300 million in July 2020.

Innovations in farming, of course, impact far more than the companies whose value rises or falls as a result. The success of even a handful of the entrepreneurs profiled here could mean significant reductions in dietary illnesses, including obesity and diabetes, and lowered health-care costs. Creative destruction, or, to use the more palatable term du jour, "disruption," could lead to the withdrawal of vast acreages from cultivation, allowing us to repurpose the American and global landscapes. On a geopolitical scale, innovation could cause today's major food exporters, such as the United States, to lose diplomatic advantage over countries that deploy modern technologies domestically.

Technological change, of course, is not deterministic, and breakthroughs can lead to vastly different social and economic outcomes. In the twentieth century, as noted by philosopher Yuval Noah Harari, "people could use the technology of the Industrial Revolution—trains, electricity, radio, telephone—in order to create communist dictatorships, fascist regimes, or liberal democracies."[8] It's encouraging, however, that food and farm innovators embrace sustainability and equity, that they are motivated by trying to feed a growing population and protect a delicate planet.

To understand the revolution afoot in agriculture, we also need to clarify "innovation." As Harold Evans put it in *They Made America*, innovation "is not simply invention; it is inventiveness put to use." Said another way, entrepreneurs seek effectiveness, not originality. Evans cites Cyrus McCormick as "not the only farmer to invent a reaper, but he was the one who initiated the financing mechanisms that made it possible for hundreds of thousands of farmers to afford the invention."[9] So while many of today's ag entrepreneurs offer advanced technological inventions, their success depends upon scaling up production and marketing to broader audiences.

In America, we have tended toward optimistic, if sometimes naive, visions of the potential of technology. Stories of entrepreneurs—such as Thomas Edison in his New Jersey laboratory and Steve Jobs in his California garage—are key parts of our culture and shared narrative. We simply enjoy upbeat visions of tomorrow.

That has been true of food, too. The February 1970 issue of *National Geographic*, for instance, offered a utopian picture of a "revolution in American agriculture." Awed by the efficiency of concentrated animal feeding operations (CAFOs)—the massive lots that fatten and prepare billions of cows, chickens, or hogs for slaughter—the magazine praised agriculture's increased productivity and technological progress.[10]

Its hopefulness, however, virtually ignored CAFOs' substantial environmental and health problems. While small farmers for years have spread manure across and nourished local fields, the concentrated facilities' massive volumes of liquefied animal wastes—which contains heavy metals and persistent hormones—leach into and poison waterways as well as emit ammonia, hydrogen sulfide, methane, and carbon dioxide. If sprayed on fields, the shit's high levels of nitrogen and phosphorus kill crops.

As modern entrepreneurs counter the magazine's 1970 embrace of industrial agriculture and its large-scale monoculture and oligopolies, we must examine their optimism. Will they move markets, have an impact at scale, reduce pollution, protect public health, and double food availability? And can those achievements be accomplished without the unintended consequences that have dogged the green revolution?

Environmentalists are often skeptical of new technologies. They cite past "advances"—think of DDT—that ended up harming the environment and human wellness. Industrial farming's "developments," as noted earlier, have damaged our land, air, and water, from soil erosion and aquatic dead zones to skyrocketing carbon emissions. Nutritionists, moreover, recall embracing margarine, thinking it an advancement that avoided the cholesterol and saturated fats in butter, only to discover that this solidified vegetable oil produces unhealthy trans fats that lead to heart attacks and cancer.

Even environmental groups that tout the importance of innovation tend to distrust entrepreneurs in favor of academics and writers who want a return to organic cultivation, cover crops, integrated livestock, and composting. To them and other "regenerative" reformers, high-tech farming seems at best a distraction. Author Jonathan Safran Foer, who eloquently implores us to give up meat products for at least breakfast and lunch, fears we won't change our behaviors "because we believe that someday, somewhere, some genius is bound to invent a miracle technology that will change our world."[11] Persuasion, however, has done little to curb meat consumption or convince farmers to embrace the additional management associated with regenerative agriculture.

Many farm-to-table chefs also distrust modernization and embrace the preindustrial agriculture of small organic farms. Yet to highlight (again) the complexity, and ferocity, of food fights, critics dub such cooks culinary elitists who serve a privileged few and fail to appreciate how technology can help provide abundant and accessible food at the scale needed to feed a growing population.[12]

David Wallace-Wells, in his insightful book *The Uninhabitable Earth*, expresses a different criticism, maintaining that "vanguard technologies," such as those featured in these chapters, will remain expensive and unavailable to "the many who are most in need."[13] Yet a growing number of ag-tech investors are betting their own money that he's wrong.

Although ag-tech entrepreneurs and regenerative reformers differ in their support for technologies and markets, they do share some common goals, particularly associated with sustainability. As noted before, more discussions among the two camps may identify complementary efforts.

Regardless of their criticisms, environmentalists and chefs are hardly the primary barriers to ag-tech innovation. Farming habits, for one, hamper change. While growers have adopted new technologies since the origin of the plow, they don't always do so quickly. Instead, they tend to adhere to convention and avoid risks. For example, regardless of the economic and environmental arguments against tillage, many growers, like their parents and grandparents, take pride in the straightness of their plow lines and relish the scent of freshly turned soil. "A lot of it is tradition," says a vineyard owner. "Bare soil, its smell and sight, is an emotional trigger for people." Another farmer admits feeling ostracized by his neighbors after he gave up plowing to reduce erosion and restore soil nutrients.[14]

Big Ag, moreover, is doing its best to maintain the status quo, particularly with expensive advertisements that encourage us to eat more processed foods made with ingredients from large farms practicing monoculture and spraying vast quantities of synthetic fertilizers and pesticides. Also significant is the government infrastructure that has grown up around industrial agriculture.

Policy makers support that infrastructure, partially in response to Big Ag's campaign contributions and lobbyists. They largely ignore industrial farming's true costs, including the damage from livestock wastes and pesticide runoff, often referred to as nonpoint-source pollution. They let corn and cattle processors enjoy cheap water, avoid sewage regulations, benefit from subsidized insurance and federal research, escape penalties for downstream dead zones, and evade regulation of greenhouse-gas emissions. One researcher calculates that if conventional food corporations lost such taxpayer benefits and paid their full production expenses, the cost of basic foods, such as a gallon of cow milk, would more than double.[15]

Public subsidies mostly support agriculture's big players. Large-scale farmers growing staples such as corn, soybeans, grains, rice, and wheat have little incentive to change practices because the federal government pays for most of their insurance against natural disasters—such as hail, drought, or floods—as well as against declines in prices of agricultural commodities.

Growers and owners of farmland annually obtain $25 billion in cash payments from the most recent farm bill.[16] In 2018 and 2019, they gained an extra $28 billion to compensate for China's refusal to buy

US crops in retaliation for the Donald Trump administration's tariffs; that's far more than was provided to the failing automotive industry during the 2008 financial crisis.[17] Acknowledging the farm lobby's political clout, economists at Kansas State University calculate that the extra government funds offered in 2018 were eight times what the trade wars actually cost farmers.[18] As the 2020 election approached, Trump showered another $46 billion on his rural base in the South and Midwest.[19]

The complex subsidy and crop-insurance programs mostly benefit large farms and agribusinesses to the detriment of small growers and high-tech start-ups. Corn cultivators and processors between 1995 and 2019 acquired $114 billion in federal payments, underwriting the promotion of snacks, beverage sweeteners, and breakfast cereals. That was four times more than for any other crop, prompting critics to call corn agriculture's welfare queen.[20] According to the Environmental Working Group, the top 10 percent of subsidy recipients grab 77 percent of the government aid, while 69 percent of farmers, mostly smaller cultivators, receive no taxpayer benefits.[21]

Also supporting Big Ag's traditionalism is a giant federal bureaucracy, the US Department of Agriculture (USDA), which spends taxpayer dollars to operate 4,500 offices and employ one hundred thousand staffers, about one for every twenty farms. Rather than explore disruptive technologies, the USDA mainly distributes subsidies to industrialized agriculture and funds land-grant universities that mostly research ways to tweak current practices and slightly enhance crop yields.

Despite decades of calls from both political parties to wean farmers from subsidies, the American Farm Bureau Federation calculates that trade bailouts, disaster insurance, insurance indemnities, and price supports accounted in 2019 for a record 40 percent of farm income.[22] Such props support the status quo, which would not survive in a free market. They also put start-up challengers, such as those profiled in this book, at a disadvantage.

Today's ag-tech entrepreneurs tend to favor transparent markets rather than government incentives; they prefer private-sector investments to politically determined subsidies. If anything, innovators hope politicians will simply stop underwriting the least healthy foods and the most polluting farm practices; protect intellectual property; use antitrust provisions to block Big Ag's oligopolistic practices; and encourage and even provide investments in rural infrastructure so farmers and ranchers, and not just city dwellers, can take advantage of digital infrastructure and connectivity advances such as 5G wireless network technology. "Bad policies are standing in the way of good food," says Eric Kessler, with Good Food Ventures. "Government needs to level the playing field and give innovators a fair chance."[23]

Regulators, moreover, could digitalize the farm sector's basic reporting to the siloed agencies overseeing air and water quality, worker safety, and permits; such moves toward a digital infrastructure would enhance trust throughout the agriculture system. Although this book's profiled innovators favor private enterprise, government policy also could stimulate farm and food creativity, with the classic models being publicly funded research that advanced semiconductors and the internet. The Netherlands provides an agriculture-focused model; two decades ago, the Dutch committed to "twice as much food using half as many resources," and this small country has become the world's second-largest exporter of food as measured by value. Concentrated around Wageningen University & Research, about fifty miles southeast of Amsterdam, is a cluster of ag-tech start-ups and experimental farms known as Food Valley.[24]

Even without government support or reform, "the sustainable food industry is on fire right now," to borrow the language of a *Forbes* article that listed a slew of recent ag-tech investments. Although farm and food innovators remain a small share of the agricultural sector, they enjoy aggressive growth. Plant-based milks, as noted earlier, rather quickly penetrated one-third of US households,[25] and one of the innovators profiled here, Jorge Heraud, optimistically observes, "When I started entrepreneuring [in 2011], entrepreneuring in agtech was a weird thing. Nobody was doing it. . . . That has completely changed now. Now every investor knows about agriculture, even if not everybody's investing yet, everybody knows."[26]

As those investments increase, so too will the range of new products and their affordability. While the initial expense of new ag tech seems daunting, it's worth remembering that research and development for the first iPhone soared to $2.6 billion, and initial microwave cookers cost $12,500. So the fact that the cell-based hamburger introduced in 2013 cost $325,000 should be no surprise, nor should the fact that entrepreneurs dropped that price two years later to under $12.[27]

The profiled innovators disagree among themselves about their appropriate relations with Big Ag. Some hope to develop partnerships that give them access to established companies' vast production and marketing resources. A few entertain the prospect of profits from Big Ag's investments or buyouts. Other entrepreneurs seek to put today's large farm and food corporations out of business.

Just as it was hard to imagine buggy-whip manufacturers disappearing, it seems hubristic of disruptors to think they will displace today's giant food and farm conglomerates. Yet businesses change with the demands of the times. And never has demand been greater for sustainable innovation. As environmental degradation and diet-related diseases become harder to ignore, consumer demand is growing for food that is healthy for both people and the planet.

Ag innovators respond to those demands with a new set of values. While industrial agriculture prioritizes yields and specialization, organic or regenerative farmers, as expressed by Wendell Berry, venerate husbandry and the oneness in all things. Today's entrepreneurs, in contrast, express support for sustainability (to protect the planet) and equity (to feed a growing population).

The environmental benefits of cutting pollution are clear, yet food and farm innovations also improve food security and safety. Many entrepreneurs combat malnutrition by delivering healthier proteins and produce, they make farming safer by curtailing poisons and reducing drudgery, and they enable growers to benefit from their harvest's special attributes. By investing in advanced technologies, innovators also spur a rural renaissance that attracts educated families and creates well-paying jobs in agricultural communities.

Food and farm innovators recognize that no single cultivation practice and no particular diet will double food availability sustainably. What they do celebrate is experimentation and the ability of science to benefit the critical agricultural sector.

The confluence of technological advances enables diverse innovation. In these pages we've met Silicon Valley scientists who deploy robots, advanced cameras, and artificial intelligence to cut the spraying of poisonous pesticides. Industrial engineers grow crops vertically in huge warehouses close to consumers. Aerospace developers use drones and sophisticated sensors to map soils and track individual plants. Biochemists produce meats in bioreactors, avoiding the slaughter of animals and slashing the methane and manure pollution from concentrated feedlots. Chefs take advantage of ignored ingredients and 3D printers to create more diverse meals. Bankers employ blockchain ledgers to trace a plant's journey, empowering consumers and helping farmers profit from producing distinctive crops.

These innovators require us to alter our sense of food, which may increasingly result from genome editing and bioreactors. They redefine growers as digitalization geeks. They change our concept of a farm to include warehouses stacked with plants that do not rest in soil or feel sunlight. They provide alternatives to degraded lands, polluted watersheds, and greenhouse gases. The offer improved nutrition, even as the world's population grows by a quarter million people each day. They force a rethinking about how we farm, what we eat, and where we obtain calories, flavor, and value.

As with any list of businesses, some highlighted in this book will fail, be acquired by larger companies, remain small, merge with complementary firms, or, quite likely, be overwhelmed by new entrepreneurs. Some of the start-up founders, moreover, will be replaced by other executives, perhaps with more marketing or financial experience. This book could have profiled different start-ups in this dynamic field, but I hope the selected entrepreneurs display the scope of disruptive transformation occurring throughout the agricultural economy.

That transformation has been astonishingly rapid. None of the profiled companies existed a single decade ago. No entrepreneur produced cell-based meat proteins without slaughtering animals. No firm challenged the horizontal farm and grew plants vertically. None plucked weeds robotically or doubled the shelf life of produce with plant-based coatings.

The simple fact that these innovations exist, in this era of troubled agriculture, brings the promise of sustainability and equity to our farms and tables.

Acknowledgments

Writing a book is an odd combination of isolation and cooperation. While I bear responsibility for the publication's contents, lots of people helped. Allow me to mention a few.

My longtime agents—Leona and Jerry Schecter—patiently reviewed early proposals, and master writer Mary Kay Zuravleff improved the book's structure and flow. Several experts—including Dan Blaustein-Reito with the Breakthrough Institute and Eric Holst with the Environmental Defense Fund—kindly reviewed sections.

Team members at Island Press have been a joy to work with. Special thanks to Emily Turner for her insightful edits, Sharis Simonian for coordinating production, Patricia Harris for copyediting, and Julie Marshall and Jen Hawse for marketing.

Special appreciation goes to Kathryn Munson for her patience, encouragement, and loving support.

Notes

Introduction. The Rise of Innovators

1. AgFunder, "2021 AgFunder AgriFoodTech Investment Report," 2021, https://agfunder.com/research/2021-AgFunder-agrifoodtech-investment-report/.

2. "Big Ag" typically refers to the largest producers of farm equipment and materials, such as fertilizers and seeds; examples are Deere & Company and Bayer. Some industry analysts label giant food processors and marketers, such as the Coca-Cola Company and Nestlé, as "Big Food." To avoid repetition, I use "Big Ag" to describe today's biggest corporations in the farm and food industries.

3. Comment by Jenette Ashtekar of CIBO, a technology platform that "scales regenerative agriculture," at the 2020 Sustainable Agriculture Summit, November 19, 2020.

4. World Commission on Environment and Development, *Our Common Future* (Oxford: Oxford University Press, 1987), p. 8.

5. McKinsey & Company, "Digital America: A Tale of the Haves and Have-Mores," December 2015, p. 5, https://www.mckinsey.com/~/media/McKinsey/Industries/Technology%20Media%20and%20Telecommunications/High%20Tech/Our%20Insights/Digital%20America%20A%20tale%20of%20the%20haves%20and%20have%20mores/MGI%20Digital%20America_Executive%20Summary_December%202015.ashx.

6. Eric Schmidt and Dror Berman, "A Super Evolution Is Coming," Medium, June 5, 2018, https://medium.com/innovationendeavors/a-super-evolution-is-coming-21b2cfad1e7.

7. Innovation Endeavors, "Accelerating the Super Evolution of Industry," n.d., accessed March 26, 2021, https://www.innovationendeavors.com/about/.

8. Seana Day, interview with the author, October 28, 2020.

9. Zoe Willingham and Andy Green, "A Fair Deal for Farmers," Center for American Progress, May 7, 2019, https://www.americanprogress.org/issues /economy/reports/2019/05/07/469385/fair-deal-farmers/.

10. Leah Nylen and Liz Crampton, "'Something Isn't Right': U.S. Probes Soaring Beef Prices," *Politico*, May 25, 2020, https://www.politico.com/news/2020 /05/25/meatpackers-prices-coronavirus-antitrust-275093#:~:text=The%20 Department%20of%20Justice%20is%20looking%20at%20the,to%20a%20 person%20with%20knowledge%20of%20the%20probe.

11. Mary Hendrickson, "The Dynamic State of Agriculture and Food: Possibilities for Rural Development?," statement at the Farm Credit Administration Symposium on Consolidation in the Farm Credit System, McLean, Virginia, February 19, 2014.

12. Pat Mooney, "Too Big to Feed: Exploring the Impacts of Mega-mergers, Consolidation, and Concentration of Power in the Agri-food Sector," International Panel of Experts on Sustainable Food Systems, October 2017, http://www.ipes-food.org/_img/upload/files/Concentration_FullReport.pdf.

13. Louisa Burwood-Taylor, "AgriFood Tech in 2018: $17bn in Funding as Biotech, Robotics, Mega Asian Deals & Record Exit Create Breakout Year," AgFunder Network, March 7, 2019, https://agfundernews.com/agrifood-tech -in-2018_17bn_funding_breakout-year.html.

14. UBS, "The Food Revolution: The Future of Food and the Challenges We Face," July 2019, https://www.ubs.com/global/en/wealth-management/chief -investment-office/sustainable-investing/2019/food-revolution/_jcr_content /mainpar/toplevelgrid_401809202/col2/teaser/linklist/actionbutton.176507 4553.file/bGluay9wYXRoPS9jb250ZW50L2RhbS9hc3NldHMvvd20vZ2xv YmFsL2Npby9kb2MvdGhlLWZvb2QtcmV2b2x1dGlvbi1qdWx5LTIwMT kucGRm/the-food-revolution-july-2019.pdf.

15. Clayton M. Christensen, *The Innovator's Dilemma* (Boston: Harvard Business Review Press, 1997).

16. Natasha Gilbert, "One-Third of Our Greenhouse Gas Emissions Come from Agriculture," *Nature*, October 31, 2012, https://www.nature.com/news/one -third-of-our-greenhouse-gas-emissions-come-from-agriculture-1.11708. According to the US Energy Information Administration, the power sector, which makes electricity, accounted for about 31 percent of greenhouse-gas emissions in 2019. See US Energy Information Administration, "How Much of U.S. Carbon Dioxide Emissions Are Associated with Electricity

Generation?," n.d., accessed March 26, 2021, https://www.eia.gov/tools/faqs
/faq.php?id=77&t=11.

17. Mark Hyman, *Food Fix: How to Save Our Health, Our Economy, Our Communities, and Our Planet—One Bite at a Time* (New York: Little, Brown Spark,
2020).

18. Chris Arsenault, "Only 60 Years of Farming Left if Soil Degradation Continues," *Scientific American*, December 5, 2014, https://www.scientificamerican
.com/article/only-60-years-of-farming-left-if-soil-degradation-continues/.

19. US Department of Agriculture, *Summary Report: 2012 National Resources
Inventory*, Natural Resources Conservation Service, Washington, DC, and
Center for Survey Statistics and Methodology, Iowa State University, August
2015, https://www.nrcs.usda.gov/Internet/FSE_DOCUMENTS/nrcseprd
396218.pdf.

20. "Pesticides" is the generic term for substances that control pests, including
insects (insecticides), weeds (herbicides), fungi (fungicides), and rodents
(rodenticides). The most common, accounting for 80 percent of pesticide use,
are herbicides used to kill or control weeds.

21. David Biello, "Oceanic Dead Zones Continue to Spread," *Scientific American*,
August 15, 2008, https://www.scientificamerican.com/article/oceanic-dead
-zones-spread/.

22. World Resources Institute et al., *World Resources Report: Creating a Sustainable
Food Future*, July 2019, https://research.wri.org/wrr-food. The Donald Trump
administration in November 2019 announced the United States would with-
draw from the Paris Agreement.

23. Mark Bittman, "How to Feed the World," *New York Times*, October 15, 2013,
https://www.nytimes.com/2013/10/15/opinion/how-to-feed-the-world.html.
About one-third of the US corn crop is converted to ethanol.

24. Kristin Ohlson, *The Soil Will Save Us: How Scientists, Farmers, and Foodies Are
Healing the Soil to Save the Planet* (New York: Rodale, 2014), p. 171.

25. Hunger Notes, "2018 World Hunger and Poverty Facts and Statistics," n.d.,
accessed March 26, 2021, https://www.worldhunger.org/world-hunger-and
-poverty-facts-and-statistics/.

26. Walter Willett et al., "Food in the Anthropocene: The EAT-*Lancet* Commission on Healthy Diets from Sustainable Food Systems," *Lancet* 393, no.
10170 (February 2, 2019): 447–492, https://doi.org/10.1016/S0140-6736
(18)31788-4.

27. Tanya Albert Henry, "Adult Obesity Rates Rise in 6 States, Exceed 35% in 7,"
American Medical Association, November 26, 2018, https://www.ama-assn

.org/delivering-care/public-health/adult-obesity-rates-rise-6-states-exceed-35-7.

28. Jordan Weissmann, "America May Have the Worst Hunger Problem of Any Rich Nation," Slate, September 4, 2014, https://slate.com/business/2014/09/american-hunger-it-s-embarrassing-by-rich-country-standards.html. The US Department of Agriculture in 2006 officially switched from the term "hunger" to "food insecurity."

29. Hank Scott, interview with the author, October 28, 2020.

30. J. D. Vivian, "Q&A with Hank Scott, President of Long and Scott Farms," *Florida Food & Farm*, June 23, 2017, http://floridafoodandfarm.com/featured/hank-scott-president-of-long-scott-farms/.

31. James M. MacDonald and Robert A. Hoppe, "Large Family Farms Continue to Dominate U.S. Agricultural Production," US Department of Agriculture, Economic Research Service, March 6, 2017, https://www.ers.usda.gov/amber-waves/2017/march/large-family-farms-continue-to-dominate-us-agricultural-production/.

32. US Department of Agriculture, Economic Research Service, "America's Diverse Family Farms," Economic Information Bulletin No. 203, December 2018, https://www.ers.usda.gov/webdocs/publications/90985/eib-203.pdf?v=9520.4.

33. American Farm Bureau Federation, "Farm Bankruptcies Rise Again," Market Intel, October 30, 2019, https://www.fb.org/market-intel/farm-bankruptcies-rise-again.

34. Hiroko Tabuchi and Nadja Popovich, "Two Biden Priorities, Climate and Inequality, Meet on Black-Owned Farms," *New York Times*, February 18, 2021, https://www.nytimes.com/2021/01/31/climate/black-farmers-discrimination-agriculture.html.

35. Vivian, "Q&A with Hank Scott."

36. Javier Cabral, "The Future of Food according to Alice Waters," *Vice*, April 5, 2016, https://www.vice.com/en_us/article/qknm8b/the-future-of-food-according-to-alice-waters.

37. Scott, interview, October 28, 2020.

38. Hyman, *Food Fix*.

39. Matthieu De Clercq, Anshu Vats, and Alvaro Biel, "Agriculture 4.0: The Future of Farming Technology," World Government Summit and Oliver Wyman, February 2018, https://www.oliverwyman.com/content/dam/oliver-wyman/v2/publications/2018/February/Oliver-Wyman-Agriculture-4.0.pdf.

40. Willett et al., "Food in the Anthropocene."

41. Damian Carrington, "New Plant-Focused Diet Would 'Transform' Planet's Future, Say Scientists," *Guardian*, January 16, 2019, https://www.theguardian .com/environment/2019/jan/16/new-plant-focused-diet-would-transform -planets-future-say-scientists.

42. Sanjeev Krishnan, "A Food System That Heals," S2G Ventures, January 2021, https://www.s2gventures.com/post/food-system-that-heals?utm_source=link edin.com&mc_cid=bd8219e429&mc_eid=d3cca6d1c4.

Part 1. Deliver Proteins

1. Winston Churchill, "Fifty Years Hence," *Popular Mechanics*, March 1932.

2. Jules B. Billard, "The Revolution in American Agriculture," *National Geographic*, February 1970.

3. Ian Sample, "Fish Fillets Grow in Tank," *New Scientist*, March 20, 2002, https://www.newscientist.com/article/dn2066-fish-fillets-grow-in-tank/.

4. Post went on to help found Mosa Meat, which seeks to commercialize cultured meat.

5. Janet Stobart, "Lab-Grown Burger from Stem Cells Introduced: Looks Good, Tastes Blah," *Los Angeles Times*, August 5, 2013, https://www.latimes.com /world/worldnow/la-fg-wn-britain-labgrown-burger-20130805-story.html.

6. Bryan Walsh, "Meat Grown from Cells Moves Out of the Lab," Axios, December 19, 2020, https://www.axios.com/future-cell-based-meat-a56be 154-c452-4af6-96d7-76ad5b9a47fb.html.

7. Elaine Watson, "How Is Coronavirus Impacting Plant-Based Meat? Impossible Foods, Lightlife, Tofurky, Meatless Farm Co, Dr. Praeger's, Weigh In," FoodNavigator-USA, April 8, 2020, https://www.foodnavigator-usa.com /Article/2020/04/06/How-is-coronavirus-impacting-plant-based-meat-Im possible-Foods-weighs-in?mc_cid=d0e7db5ad1&mc_eid=1a60d2836e#.

8. Brian Kateman, "Plant-Based Meat Has Officially Reached 'Global Phenomenon' Status," *Entrepreneur*, February 26, 2020, https://www.entrepreneur .com/article/346116.

9. Kelsey Piper, "The Rise of Meatless Meat, Explained," Vox, February 20, 2020, https://www.vox.com/2019/5/28/18626859/meatless-meat-explained-vegan -impossible-burger.

10. Laura Reiley, "Road to Recovery: Going It Alone in Two of America's Agricultural Towns," *Washington Post*, December 8, 2020, https://www.wash ingtonpost.com/graphics/2020/road-to-recovery/farmers-ranchers-corona virus-food-california-west-virginia/.

11. Larissa Zimberoff, "Fake-Meat Startups Rake In Cash amid Food Supply

Worries," Bloomberg, May 1, 2020, https://www.bloomberg.com/news
/articles/2020-05-01/fake-meat-startups-rake-in-cash-amid-coronavirus
-food-worries.

12. Thomas Franck, "Alternative Meat to Become $140 Billion Industry in a
Decade, Barclays Predicts," CNBC, May 23, 2019, https://www.cnbc
.com/2019/05/23/alternative-meat-to-become-140-billion-industry-barclays
-says.html.

13. Jo Warrick, "They Die Piece by Piece," *Washington Post*, April 10, 2001,
https://www.washingtonpost.com/archive/politics/2001/04/10/they-die-piece
-by-piece/f172dd3c-0383-49f8-b6d8-347e04b68da1/.

14. "The Jobs We Need," editorial, *New York Times*, July 5, 2020.

15. Yuval Noah Harari, foreword to *Clean Meat: How Growing Meat without
Animals Will Revolutionize Dinner and the World*, by Paul Shapiro (New York:
Gallery Books, 2018), p. x.

16. Jonathan Foley, "A Five-Step Plan to Feed the World," *National Geographic*,
March 4, 2015, https://www.nationalgeographic.com/foodfeatures/feeding
-9-billion/.

17. "Food Choices and the Planet," EarthSave, n.d., accessed March 26, 2021,
http://www.earthsave.org/environment.htm.

18. Robert Goodland and Jeff Anhang, "Livestock and Climate Change," *World
Watch Magazine*, November/December 2009, https:awellfedworld.org
/wp-content/uploads/Livestock-Climate-Change-Anhang-Goodland.pdf; the
authors defend the 51 percent estimate. By not including the impact of fertilizers used to grow food for livestock and a couple of other factors, the *New
York Times* uses 14.5–18 percent; see Lisa Friedman, Kendra Pierre-Louis, and
Somini Sengupta, "The Meat Question, by the Numbers," *New York Times*,
January 25, 2018, https://www.nytimes.com/2018/01/25/climate/cows
-global-warming.html. The beef industry in 2020 funded its own study saying that US cattle account for only 3 percent of man-made greenhouse gases.

19. Environmental Defense Fund, "Methane: The Other Important Greenhouse
Gas," n.d., accessed March 26, 2021, https://www.edf.org/climate/methane
-other-important-greenhouse-gas.

20. Claire Felter, "The End of Antibiotics?," Council on Foreign Relations,
November 15, 2019, https://www.cfr.org/backgrounder/end-antibiotics.

21. Food and Agriculture Organization of the United Nations, "Livestock's Long
Shadow: Environmental Issues and Options," 2006, http://www.fao.org/3/a
0701e/a0701e00.htm.

22. Bill Maher, "A-hole in One Shouldn't Be Obama's Game," HuffPost, May 25,

2011, http://www.huffingtonpost.com/bill-maher/new-rule-a-hole-in-one-sh _b_259281.html.

23. "Global Meat-Eating Is on the Rise, Bringing Surprising Benefits," *Economist*, May 4, 2019, https://www.economist.com/international/2019/05/04/global -meat-eating-is-on-the-rise-bringing-surprising-benefits.

24. Mikael Djanian and Nelson Ferreira, "Agriculture Sector: Preparing for Disruption in the Food Value Chain," McKinsey & Company, April 2020, https://www.mckinsey.com/~/media/McKinsey/Industries/Agriculture/Our %20Insights/Agriculture%20sector%20Preparing%20for%20disruption%20 in%20the%20food%20value%20chain/Agriculture-sector-Preparing-for -disruption-in-the-food-value-chain-vF2.pdf.

25. Nicholas Bakalar, "Eat Less Meat, Live Longer?," *New York Times*, April 11, 2019, https://www.nytimes.com/2019/04/11/well/eat/eat-less-meat-live-long er.html.

26. Aaron E. Carroll, "The Real Problem with Beef," *New York Times*, October 1, 2019, https://www.nytimes.com/2019/10/01/upshot/beef-health-climate -impact.html.

27. Kathy Freston, "E. Coli, Salmonella and Other Deadly Bacteria and Pathogens in Food: Factory Farms Are the Reason," HuffPost, March 18, 2010, https://www.huffpost.com/entry/e-coli-salmonella-and-oth_b_415240.

28. John A. Painter et al., "Attribution of Foodborne Illnesses, Hospitalizations, and Deaths to Food Commodities by Using Outbreak Data, United States, 1998–2008," *Emerging Infectious Diseases* 19, no. 3 (March 2013): 407–415, https://dx.doi.org/10.3201/eid1903.111866.

29. Jayson Lusk, *Unnaturally Delicious: How Science and Technology Are Serving Up Super Foods to Save the World* (New York: St. Martin's Press, 2016), p. 111.

30. It takes the manure from one or two cows to fertilize an acre, which is about the size of a football field.

31. Tad Friend, "Can a Burger Help Solve Climate Change?," *New Yorker*, September 23, 2019, https://www.newyorker.com/magazine/2019/09/30/can -a-burger-help-solve-climate-change.

Chapter 1. Josh Tetrick, Eat Just—Rethinking the Chicken and the Egg

1. Jeff Fromm, "Let Your Purpose Be Your Driver: Josh Tetrick and JUST," *Forbes*, November 21, 2019, https://www.forbes.com/sites/jefffromm/2019/11/21 /let-your-purpose-be-your-driver-josh-tetrick-and-just-egg/#15c595c47eb4.

2. Paul Shapiro, *Clean Meat: How Growing Meat without Animals Will Revolutionize Dinner and the World* (New York: Gallery Books, 2018), p. 160.

3. Fromm, "Let Your Purpose Be Your Driver."

4. Chase Purdy, *Billion Dollar Burger: Inside Big Tech's Race for the Future of Food* (New York: Portfolio/Penguin, 2020), p. 77.

5. Tetrick cofounded the firm with Josh Balk, his childhood best friend, fellow vegan activist, and executive at the Humane Society of the United States. Although Balk provides frequent guidance, he holds no equity in the company. Erica Swallow, "Hampton Creek's Plan to Reimagine the Future of Food," Mashable, August 27, 2014, https://mashable.com/2014/08/27/hampton-creek/.

6. Biana Bosker, "Mayonnaise, Disrupted," *Atlantic*, November 2017, https://www.theatlantic.com/magzine/archive/2017/11/hampton-creek-josh-tetrick-mayo-mogul/540642/.

7. Bosker, "Mayonnaise, Disrupted."

8. Mary Ellen Shoup, "Eat Just Inc. Closes $200m Funding Round: 'You Can Eat an Animal without Slaughter. I Want to Build a Brand around That,'" FoodNavigator-USA, March 24, 2021, https://www.foodnavigator-usa.com/Article/2021/03/24/Eat-Just-Inc.-closes-200m-funding-round-You-can-eat-an-animal-without-slaughter.-I-want-to-build-a-brand-around-that.#.

9. Bosker, "Mayonnaise, Disrupted."

10. Just Egg, "Meet Ian & Learn about Cultured Meat," Medium, August 25, 2017, https://medium.com/eatjust/clean-meat-feeding-the-future-2636077ed906.

11. United Nations Environment Programme, "Preventing the Next Pandemic—Zoonotic Diseases and How to Break the Chain of Transmission," July 6, 2020, https://www.unep.org/resources/report/preventing-future-zoonotic-disease-outbreaks-protecting-environment-animals-and.

12. Jennifer Marston, "A Conversation with Josh Tetrick of JUST," Spoon, August 20, 2020, https://thespoon.tech/video-50-million-plant-based-eggs-later-just-keeps-innovating/.

13. Marston, "Conversation with Josh Tetrick."

14. Eillie Anzilotti, "Plant-Based Eggs Are Coming for Your Breakfast Sandwiches," Fast Company, August 3, 2018, https://www.fastcompany.com/90212355/plant-based-eggs-are-coming-for-your-breakfast-sandwiches.

15. Elaine Watson, "Eat Just Aims to Achieve Operational Profitability by End of 2021, Mulls IPO: 'Our Addressable Market Is the $238bn Global Chicken Egg Market,'" FoodNavigator-USA, August 20, 2020, https://www.foodnavigator-usa.com/Article/2020/08/20/Eat-Just-aims-to-achieve-operational-profitability-by-end-of-2021-mulls-IPO.

16. Watson, "Eat Just Aims to Achieve Operational Profitability."

17. Shoup, "Eat Just Inc. Closes $200m Funding Round."

18. Mary Ellen Shoup, "Whole Foods CEO and Eat Just CEO Talk the Future of Food and Importance of Innovation," FoodNavigator-USA, October 5, 2020, https://www.foodnavigator-usa.com/Article/2020/10/05/Whole-Foods-CEO -and-Eat-Just-CEO-talk-the-future-of-food-and-importance-of-innovation.

19. Sally Ho, "Eat Just Makes First-Ever Commercial Sale of Cultured Meat in Singapore," green queen, December 16, 2020, https://www.greenqueen.com .hk/eat-just-makes-first-ever-commercial-sale-of-cultured-meat-in-singapore/.

20. Mike Ives, "In Global First, Lab-Grown Meat Is Approved for Sale in Singapore," *New York Times*, December 3, 2020, p. B5.

21. Eat Just's website, accessed March 25, 2021, https://www.ju.st/stories/meat.

22. Adele Peters, "The Meat Growing in This San Francisco Lab Will Soon Be Available at Restaurants," Fast Company, December 11, 2018, https://www .fastcompany.com/90278853/the-meat-growing-in-this-san-francisco-lab-will -soon-be-available-at-restaurants.

23. Chase Purdy, "The Board of Silicon Valley's Biggest Food Tech Company Has Basically Called It Quits," Quartz, July 17, 2017, https://qz.com/1031431 /hampton-creek-board-members-quit-leaving-ceo-josh-tetrick-as-the-vegan -mayo-companys-sole-director/.

24. Sam Thielman and Dominic Rushe, "Government-Based Egg Lobby Tried to Crack Food Startup, Emails Show," *Guardian*, September 2, 2015, https:// www.theguardian.com/us-news/2015/sep/02/usda-american-egg-board -hampton-creek-just-mayo.

25. Peters, "Meat Growing in This San Francisco Lab."

26. Amanda Weston, "Eat Just's 'No-Kill' Meat Will 'Change the Food System,' Says CEO," Cheddar, December 14, 2020, https://cheddar.com/media/eat -justs-no-kill-meat-will-change-the-food-system-says-ceo.

27. Jenny Kleeman, *Sex Robots and Vegan Meat: Adventures at the Frontier of Birth, Food, Sex, and Death* (New York: Pegasus Books, 2020).

28. Jonathan Ho, "Social Entrepreneur Josh Tetrick of Eat Just Talks Sustainability and the Future of Food," Augustman, February 25, 2021, https://www .augustman.com/sg/amselect/august-mentor/social-entrepreneur-josh-tetrick -of-eat-just-talks-sustainability-and-the-future-of-food/.

29. Ho, "Social Entrepreneur Josh Tetrick."

30. Shoup, "Whole Foods CEO and Eat Just CEO Talk the Future of Food."

Chapter 2. Uma Valeti, UPSIDE Foods—Avoiding Animal Slaughter

1. Jeff Bercovici, "Why This Cardiologist Is Betting That His Lab-Grown Meat

Startup Can Solve the Global Food Crisis," *Inc.*, November 2017, https://
www.inc.com/magazine/201711/jeff-bercovici/memphis-meats-lab-grown
-meat-startup.html.

2. Paul Shapiro, *Clean Meat: How Growing Meat without Animals Will Revolution-ize Dinner and the World* (New York: Gallery Books, 2018), p. 113.

3. Bercovici, "Why This Cardiologist Is Betting."

4. Mary Allen, "How a Cardiologist Is Using Meat to Save More Lives," Good Food Institute, August 10, 2018, https://www.gfi.org/how-a-cardiologist-is
-using-meat-to-save.

5. Bercovici, "Why This Cardiologist Is Betting."

6. Allen, "How a Cardiologist Is Using Meat to Save More Lives."

7. "Study Says 48% of Chicken Has E. Coli, Some Question Results," HuffPost, April 12, 2012, https://www.huffpost.com/entry/chicken-e-coli_n_1420393.

8. Jonathan Foley, "A Five-Step Plan to Feed the World," *National Geographic*, March 4, 2015, https://www.nationalgeographic.com/foodfeatures/feeding
-9-billion/.

9. Peter Singer, *Animal Liberation: A New Ethics for Our Treatment of Animals* (New York: HarperCollins, 1975).

10. The European, "Peter Singer: 'We Need to Be More Cautious with Our Cau-tion," HuffPost, October 30, 2014, https://www.huffpost.com/entry/peter
-singer-cautious-caution_n_6075116.

11. Uma Valeti, interview with the author, February 5, 2021.

12. Bercovici, "Why This Cardiologist Is Betting."

13. Aleksandra Sagan, "Cultured Meat, Milk and Eggs Poised to Enter Grocery Stores," CTV News, March 10, 2016, https://www.ctvnews.ca/sci-tech
/cultured-meat-milk-and-eggs-poised-to-enter-grocery-stores-1.2811263.

14. Uma Valeti, interview with the author, November 30, 2020.

15. Hanna L. Tuomisto and M. Joost Teixeira de Mattos, "Environmental Impacts of Cultured Meat Production," *Environmental Science & Technology* 45, no. 14 (2011): 6117–6123, https://doi.org/10.1021/es200130u.

16. Kadence International, "Only 20% of U.S. Adults Likely to Buy 'Clean Meat,'" Deli Market News, October 23, 2018, https://www.delimarketnews
.com/press-release/only-20-us-adults-likely-buy-clean-meat. A separate poll by the industry-supported Cellular Agriculture Society found that 77 percent of college students would "probably" or "definitely" eat cultured meat after being told of its environmental, health, and ethical benefits.

17. Mary Catherine O'Connor, "Lab-Grown Meat, Part 1: Give Cells a Chance,"

Climate Confidential, September 30, 2014, https://climateconfidential.com
/portfolio/the-future-of-lab-grown-meat-part-1-give-cells-a-chance/.

18. Valeti, interview, November 30, 2020.

19. Chase Purdy, *Billion Dollar Burger: Inside Big Tech's Race for the Future of Food*
(New York: Portfolio/Penguin, 2020), p. 133.

20. Commissioner of Food and Drugs, US Food and Drug Administration,
"Statement from USDA Secretary Perdue and FDA Commissioner Gottlieb
on the Regulation of Cell-Cultured Food Products from Cell Lines of Livestock and Poultry," November 16, 2018, https://www.fda.gov/news-events
/press-announcements/statement-usda-secretary-perdue-and-fda-commission
er-gottlieb-regulation-cell-cultured-food-products.

21. Bryan Walsh, "Meat Grown from Cells Moves Out of the Lab," Axios, December 19, 2020, https://www.axios.com/future-cell-based-meat-a56be154-c452
-4af6-96d7-76ad5b9a47fb.html.

22. MarketsandMarkets, "Cultured Meat Market by Source (Poultry, Beef, Seafood, Pork, and Duck), End-Use (Nuggets, Burgers, Meatballs, Sausages, Hot
Dogs), and Region (North America, Europe, Asia Pacific, Middle East and
Africa, South America)—Global Forecast to 2032," accessed March 25, 2021,
https://www.marketsandmarkets.com/Market-Reports/cultured-meat-market
-204524444.html.

23. Carsten Gerhardt et al., "How Will Cultured Meat and Meat Alternatives
Disrupt the Agricultural and Food Industry?," Kearney, 2019, https://www
.kearney.com/documents/20152/2795757/How+Will+Cultured+Meat+and
+Meat+Alternatives+Disrupt+the+Agricultural+and+Food+Industry.pdf/06ec
385b-63a1-71d2-c081-51c07ab88ad1?t=1559860712714.

24. Shapiro, *Clean Meat*, p. 80.

25. Jade Scipioni, "Tyson Foods CEO: The Future of Food Might Be Meatless,"
FOX Business, March 7, 2017, https://www.foxbusiness.com/features/tyson
-foods-ceo-the-future-of-food-might-be-meatless.

26. Valeti, interview, February 5, 2021.

27. Bercovici, "Why This Cardiologist Is Betting."

28. Mary Ellen Shoup, "Whole Foods CEO and Eat Just CEO Talk the Future of
Food and Importance of Innovation," FoodNavigator-USA, October 5, 2020,
https://www.foodnavigator-usa.com/Article/2020/10/05/Whole-Foods-CEO
-and-Eat-Just-CEO-talk-the-future-of-food-and-importance-of-innovation.

29. Valeti, interview with the author, November 30, 2020.

30. Allen, "How a Cardiologist Is Using Meat to Save More Lives."

31. Catherine Tubb and Tony Seba, "Rethinking Food and Agriculture, 2020–

2030," September 2019, RethinkX, https://www.rethinkx.com/food-and
-agriculture.

32. Avery Ellfeldt, "Can Lab-Grown Meat Save the Planet—and Dinner?," E&E
News, May 15, 2020, https://www.eenews.net/stories/1063136269.

33. Benjamin Aldes Wurgaft, *Meat Planet: Artificial Flesh and the Future of Food*
(Oakland: University of California Press, 2019), p. 123.

34. Bianca Phillips, "Meet the New Meat," *Memphis Flyer*, March 3, 2016, https:
//www.memphisflyer.com/memphis/meet-the-new-meat/Content?oid=449
4531.

35. Valeti, interview, November 30, 2020.

Chapter 3. Patrick Brown, Impossible Foods—Making Burgers from Plants

1. Pamela C. Ronald and Raoul W. Adamchak, *Tomorrow's Table: Organic
Farming, Genetics, and the Future of Food* (New York: Oxford University Press,
2008), p. 243.

2. Tad Friend, "Can a Burger Help Solve Climate Change?," *New Yorker*, September 23, 2019, https://www.newyorker.com/magazine/2019/09/30/can-a
-burger-help-solve-climate-change.

3. Bob Woods, "Inside Impossible Foods' Mission to Mass-Produce the Fake
Burger of the Future," CNBC, May 15, 2019, https://www.cnbc.com/2019
/05/14/impossible-foods-racing-to-mass-produce-fake-burger-of-the-future
.html.

4. Danielle Centoni, "An Honest Review of the Cult-Favorite Impossible
Burger," kitchn, June 6, 2019, https://www.thekitchn.com/impossible-burg
er-22903550#:~:text=About%20a%20year%20ago%2C%20when%20
chef%20Sarah%20Schafer,a%20vegan%2C%20plant-based%20patty%20
that%20%E2%80%9Cbleeds%2C%E2%80%9D%20she%20scoffed.

5. Mary Ellen Shoup, "'We Are Meat': Impossible Foods Launches First
National Ad Campaign Targeting Meat Eaters," FoodNavigator-USA, April
6, 2021, https://www.foodnavigator-usa.com/Article/2021/04/06/We-are
-meat-Impossible-Foods-launches-first-national-ad-campaign-targeting-meat
-eaters?utm_source=newsletter_daily&utm_medium=email&utm_campaign
=06-Apr-2021.

6. Jordan Valinsky and Danielle Wiener-Bronner, "America Is Running Out of
Impossible Burgers," CNN Business, April 30, 2019, https://www.cnn.com
/2019/04/30/business/impossible-meat-shortage/index.html.

7. Richard Martyn-Hemphill, "Fermentation Sees Foodtech Investors Bubbling
Up in 2020," AgFunder Network, October 5, 2020, https://agfundernews.com
/fermentation-sees-foodtech-investors-bubbling-up-in-2020.html#:~:text=Fer

mentation%20%E2%80%94%20which%20involves%20working%20with%
20microbes%20like,and%20%E2%80%98bloodiness%E2%80%99%20of%
20beef%20in%20its%20plant-based%20patties.

8. Elaine Watson, "Impossible Foods: 'Our Goal Is to Produce a Full Range of
Meats and Dairy Products for Every Cultural Region in the World,'" Food-
Navigator-USA, April 12, 2018, https://www.foodnavigator-usa.com/Article
/2018/04/13/Impossible-Foods-Our-goal-is-to-produce-a-full-range-of-meats
-and-dairy-products-for-every-cultural-region-in-the-world.

9. Friend, "Can a Burger Help Solve Climate Change?"

10. David Lee, interview with the author, June 18, 2020.

11. Woods, "Inside Impossible Foods' Mission."

12. Joseph Hincks, "Meet the Founder of Impossible Foods, Whose Meat-Free
Burgers Could Transform the Way We Eat," *Time*, April 23, 2018, https://
time.com/5247858/impossible-foods-meat-plant-based-agriculture/.

13. Hincks, "Meet the Founder of Impossible Foods."

14. Friends of the Earth, "'Bleeding' Veggie Burger Has 'No Basis for Safety,'
According to FDA," August 8, 2017, https://foe.org/news/2017-08-bleeding
-veggie-burger-has-no-basis-for-safety-according-to-fda/.

15. Friend, "Can a Burger Help Solve Climate Change?"

16. Anahad O'Connor, "Fake Meat vs. Real Meat," *New York Times*, December 3,
2019, https://www.nytimes.com/2019/12/03/well/eat/fake-meat-vs-real-meat
.html.

17. Aaron E. Carroll, "The Real Problem with Beef," *New York Times*, October 1,
2019, https://www.nytimes.com/2019/10/01/upshot/beef-health-climate
-impact.html.

18. Beyond Meat's CEO is Ethan Brown, no relation to Impossible Foods' Pat
Brown.

19. Tyson Foods in 2019 exited its 6.5 percent position in plant-protein-focused
Beyond Meat but then introduced its own alternative-meat product.

20. Kate Uhlman, "Silicon Valley's Next Disruption: The Beef Industry," Penin-
sula Press, April 10, 2018, http://peninsulapress.com/2018/04/10/silicon
-valleys-next-disruption-the-beef-industry/.

21. Uma Valeti, interview with the author, February 5, 2021.

22. Elaine Watson, "Impossible Foods Raises Another $200m; Products under
Development Include Milk and Steaks," FoodNavigator-USA, August 14,
2020, https://www.foodnavigator-usa.com/Article/2020/08/14/Impossible
-Foods-raises-another-200m-products-under-development-include-milk

-and-steaks?utm_source=newsletter_daily&utm_medium=email&utm_cam
paign=14-Aug-2020.

23. Friend, "Can a Burger Help Solve Climate Change?"

24. O'Connor, "Fake Meat vs. Real Meat."

25. Sully Barrett, "Beyond Meat, Impossible Foods Face a New 'Fake Meat' Foe
with Long, Controversial History," CNBC, February 6, 2020, https://www
.cnbc.com/2020/02/06/beyond-meat-impossible-foods-face-new-powerful
-fake-meat-foe.html?mc_cid=e0b25a19a6&mc_eid=1a60d2836e.

26. Shoup, "'We Are Meat.'"

27. UBS, "The Food Revolution: The Future of Food and the Challenges We
Face," July 2019, https://www.ubs.com/global/en/wealth-management/chief
-investment-office/sustainable-investing/2019/food-revolution/_jcr_content
/mainpar/toplevelgrid_401809202/col2/teaser/linklist/actionbutton.176507
4553.file/bGluay9wYXRoPS9jb250ZW50L2RhbS9hc9hc3NldHMvd20vZ2x
vYmFsL2Npby9kb2MvdGhlLWZvb2QtctcmV2b2x1dGlvbi1qdWx5LTIw
MTkucGRm/the-food-revolution-july-2019.pdf.

28. Patrick Brown, interview with the author, November 20, 2020.

Chapter 4. James Corwell, Ocean Hugger Foods—Turning Tomatoes into Tuna

1. Ransom A. Myers and Boris Worm, "Rapid Worldwide Depletion of Preda-
tory Fish Communities," *Nature* 423 (May 15, 2003): 280–283, https:
//doi.org/10.1038/nature01610.

2. DNews, "Oceans' Fish Could Disappear by 2050," Seeker, May 17, 2010,
https://www.seeker.com/oceans-fish-could-disappear-by-2050-1765058733
.html.

3. James Corwell, interview with the author, September 26, 2019.

4. Plantbased TourGuide, podcast with Chef James Corwell, Stitcher, September 17, 2017, https://www.stitcher.com/show/vegan-tourguide/episode/vtg
-037-chef-james-corwell-51512933.

5. Corwell, interview, September 26, 2019.

6. Corwell, interview, September 26, 2019.

7. James Corwell, interview with the author, December 10, 2020.

8. Corwell, interview, September 26, 2019.

9. Sharon Palmer, "Plant Chat, James Corwell, Tomato Sushi," sharonpalmer
.com, December 3, 2014, https://sharonpalmer.com/2014-12-03-in-the
-studio-with-sharon-james-corwell-tomato-sushi/.

10. Katrina Fox, "Vegan Seafood Is About to Become Big Business—and Not a

Moment Too Soon," *Forbes*, August 6, 2018, https://www.forbes.com/sites
/katrinafox/2018/08/06/vegan-seafood-is-about-to-become-big-business-and
-not-a-moment-too-soon/#5ff8c2fe645d.

11. Corwell, interview, September 26, 2019.

12. Joshua Spodek, "How Your Tuna Is About to Get Plant-Based," *Inc.*, October
31, 2018, https://www.inc.com/joshua-spodek/how-your-tuna-is-about-to
-get-sustainable.html.

13. Palmer, "Plant Chat."

14. Spodek, "Your Tuna."

15. Spodek, "Your Tuna."

16. Plantbased TourGuide, podcast with Chef James Corwell.

Chapter 5. Virginia Emery, Beta Hatch—Farming Insects

1. Comment by Patrick McGuire, managing partner at Action Mary, August 2,
2017. Posted on Virginia Emery's LinkedIn page.

2. "Insect" and "bug" often are used interchangeably, but from an entomolog-
ical view they are different. Bugs are insects, yet not all insects are bugs. See
http://www.naturallynorthidaho.com/2014/09/whats-difference-between-bug
-and-beetle.htm.

3. Virginia Emery, interview with the author, September 10, 2019.

4. Emery, interview, September 10, 2019.

5. Emery, interview, September 10, 2019.

6. Kara Carlson, "Wriggle Room: Beta Hatch's Insect Farm Grows Millions of
Mealworms," *Seattle Times*, June 22, 2017, https://www.seattletimes.com/busi
ness/wiggle-room-beta-hatchs-insect-farm-grows-millions-of-mealworms/.

7. See Beta Hatch's website, https://betahatch.com.

8. Emery, interview, September 10, 2019.

9. Emery, interview, September 10, 2019.

10. Lauren Manning, "Why Insect Protein Startup Beta Hatch Is Swapping Its
Seattle Home Base for a Rural Town in Washington," AgFunder Network,
May 7, 2019, https://agfundernews.com/breaking-why-insect-protein-start
up-beta-hatch-is-swapping-its-Seattle-home-base-for-a-rural-town-in-wash
ington.html.

11. Kurt Schlosser, "Insect-Growing Startup Beta Hatch Raises $3M to Further
Innovate and Build Giant New Facility," GeekWire, May 21, 2020, https://
www.geekwire.com/2020/insect-growing-startup-beta-hatch-raises-3m-inno
vate-build-giant-new-facility/.

12. AgFunder Network, "AgriFood Tech Investing Report: Year in Review 2018," https://agfunder.com/research/agrifood-tech-investing-report-2018/.

13. Susan Adams, "A Seattle Startup, Beta Hatch, Thinks Growing Bugs for Animal Feed Is a Billion-Dollar Opportunity," *Forbes*, July 6, 2017, https://www.forbes.com/sites/forbestreptalks/2017/07/06/a-seattle-startup-beta-hatch-thinks-growing-bugs-for-animal-feed-is-a-billion-dollar-opportunity/?sh=7f7d8afe3467.

14. Manning, "Insect Protein Startup Beta Hatch."

15. Emery, interview, September 10, 2019.

16. Adams, "Seattle Startup, Beta Hatch."

17. Emery, interview, September 10, 2019.

Chapter 6. Leonard Lerer, Back of the Yards Algae Sciences—Growing Algae and Mycelia

1. Leonard Lerer, interview with the author, September 26, 2019.

2. Leonard Lerer, interview with the author, November 18, 2020.

3. Lerer, interview, November 18, 2020.

4. Akshat Rathi, "The Revolutionary Technology Pushing Sweden toward the Seemingly Impossible Goal of Zero Emissions," Quartz, June 21, 2017, https://qz.com/1010273/the-algoland-carbon-capture-project-in-sweden-uses-algae-to-help-the-country-reach-zero-emissions/.

5. Lerer, interview, September 26, 2019.

6. Warren Belasco, "Algae Burgers for a Hungry World? The Rise and Fall of Chlorella Cuisine," *Technology and Culture* 38, no. 3 (July 1997): 608–634, https://doi.org/10.2307/3106856.

7. Lerer, interview, September 26, 2019.

8. Joe Gan, "AFN Introduces . . . SinGENE and Back of the Yards Algae Sciences," AgFunder Network, October 21, 2019, https://agfundernews.com/afn-introduces-singene-and-back-of-the-yards-algae-sciences.html.

9. Eduardo Garcia, "Where's the Waste? A 'Circular' Food Economy Could Combat Climate Change," *New York Times*, September 24, 2019, https://www.nytimes.com/2019/09/21/climate/circular-food-economy-sustainable.html.

10. Jane Byrne, "Algae as Cheap Cattle Feed Supplement: University of Queensland Shows How It Is Done," FeedNavigator.com, September 1, 2014, https://www.feednavigator.com/Article/2014/09/02/Aus-team-develops-processs-to-produce-low-cost-algae-for-cattle-feed.

11. Matthieu De Clercq, Anshu Vats, and Alvaro Biel, "Agriculture 4.0: The

Future of Farming Technology," World Government Summit and Oliver Wyman, February 2018, https://www.oliverwyman.com/content/dam/oliver -wyman/v2/publications/2018/February/Oliver-Wyman-Agriculture-4.0.pdf.

12. Leonard Lerer and Cedric Kamaleson, "Growth, Yield, and Quality in Hydroponic Vertical Farming—Effects of Phycocyanin-Rich Spirulina Extract," *Preprints* 2020, 2020110354.

13. Lerer, interview, November 18, 2020.

14. Lerer, interview, November 18, 2020.

Part 2. Reduce Food Waste

1. Quentin Fottrell, "Food for Thought This Thanksgiving: 40% of Groceries Are Thrown Out Every Year," MarketWatch, November 21, 2018, https:// www.marketwatch.com/story/this-is-why-americans-throw-out-165-billion -in-food-every-year-2016-07-22.

2. RTS, "Food Waste in America in 2021," https://www.rts.com/resources /guides/food-waste-america/. Seana Day of The Mixing Bowl calls US food waste "a $1 trillion question"; see Seana Day, "Agtech Landscape: Tracking 1,600+ Startups Innovating on the Farm and in the Supply Chain," *Forbes*, September 3, 2019, https://www.forbes.com/sites/themixingbowl/2019/09 /03/agtech-landscape-tracking-1600-startups-innovating-on-the-farm-and-in -the-supply-chain/#7bcc8e43b62d.

3. Dana Gunders, "Wasted: How America Is Losing up to 40 Percent of Its Food from Farm to Fork to Landfill," Natural Resources Defense Council, August 16, 2017, https://www.nrdc.org/resources/wasted-how-america -losing-40-percent-its-food-farm-fork-landfill.

4. Zach Conrad et al., "Relationship between Food Waste, Diet Quality, and Environmental Sustainability," *PLoS ONE* 13, no. 4 (April 18, 2018): e0195405, https://doi.org/10.1371/journal.pone.0195405.

5. Gunders, "Wasted."

6. Food and Agriculture Organization of the United Nations, "Food Wastage Footprint: Impacts on Natural Resources; Summary Report," 2013, http:// www.fao.org/3/i3347e/i3347e.pdf.

7. Paul Hawken, ed., *Drawdown: The Most Comprehensive Plan Ever Proposed to Reverse Global Warming* (New York: Penguin Books, 2017); Project Drawdown, "Table of Solutions," https://www.drawdown.org/solutions-summary-by -rank.

8. Monica Eng, "Most Produce Loses 30 Percent of Nutrients Three Days after Harvest," *Chicago Tribune*, July 10, 2013, https://www.chicagotribune.com

/dining/ct-xpm-2013-07-10-chi-most-produce-loses-30-percent-of-nutrients
-three-days-after-harvest-20130710-story.html.

9. Audre Kapacinskas, "S2G Ventures on the Future of Food Retail," Food +
Tech Connect, November 29, 2020, https://foodtechconnect.com/2020/11
/29/s2g-ventures-on-the-future-of-food-retail/. S2G Ventures, "The Future
of Food: Through the Lens of Retail," 2020.

10. Rivka Galchen, "How South Korea Is Composting Its Way to Sustainability,"
New Yorker, March 2, 2020, https://www.newyorker.com/magazine/2020/03
/09/how-south-korea-is-composting-its-way-to-sustainability.

11. US Department of Agriculture, Agricultural Marketing Service, "National
Count of Farmers Market Directory Listings," August 2019, https://www
.ams.usda.gov/sites/default/files/media/NationalCountofFarmersMarket
DirectoryListings082019.pdf.

12. Jodi Helmer, "Why Are So Many Farmers Markets Failing? Because the
Market Is Saturated," NPR, March 17, 2019, https://www.npr.org/sections
/thesalt/2019/03/17/700715793/why-are-so-many-farmers-markets-failing
-because-the-market-is-saturated.

13. FMI—The Food Industry Association, "FMI, Nielsen Study Shows Online
Grocery Forecast Increase to $143B by 2025," *Food Logistics*, February 6,
2020, https://www.foodlogistics.com/warehousing/article/21114583/fmi-the
-food-industry-association-fmi-nielsen-study-shows-online-grocery-forecast
-increase-to-143b-by-2025.

14. Day, "Agtech Landscape."

**Chapter 7. Irving Fain, Bowery Farming—Bringing Crops Closer
to Consumers**

1. Irving Fain, interview with the author, December 5, 2021.

2. "Irving Fain," *The Proof*, no. 19, https://www.theproofwellness.com/irving
-fain.

3. Daniel Lipson, "This Vertical Farming Startup Is Valued at $27.5 Million,"
iGrow, August 23, 2017, https://www.igrow.news/igrownews/this-vertical
-farming-startup-is-valued-at-275-million#:~:text=CEO%20Irving%20Fain
%20describes%20his%20company%20as%20a,help%20curb%20the%20
environmental%20impact%20of%20the%20industry.

4. Mary Ann Azevedo, "Bowery Raises $90M Series B for Pesticide-Free Farm-
ing," Crunchbase News, December 12, 2018, https://news.crunchbase.com
/news/bowery-raises-90m-series-b-for-pesticide-free-farming/.

5. Louisa Burwood-Taylor, "Bowery CEO: Building Tech Is Expensive, Takes

Time but Has Direct Impact on Economics of Vertical Farming," AgFunder Network, November 12, 2019, https://agfundernews.com/bowery-ceo-build ing-tech-is-expensive-takes-time-but-has-direct-impact-on-economics-of -vertical-farming.html.

6. John Seabrook, "Machine Hands," *New Yorker*, April 15, 2019, p. 55.

7. Walter Robb, former co-CEO of Whole Foods Market, S2G Ventures pod- cast, December 8, 2020.

8. Irving Fain, "Storm at the Door: Climate Week 2019 and the Road Ahead," Medium, September 23, 2019, https://news.boweryfarming.com/storm-at-the -door-climate-week-2019-and-the-road-ahead-4a447f3807f.

9. Stan Cox, "The Vertical Farming Scam," CounterPunch, December 11, 2012, https://www.counterpunch.org/2012/12/11/the-vertical-farming-scam/.

Chapter 8. James Rogers, Jenny Du, and Louis Perez, Apeel Sciences— Coating Foods

1. James Rogers, Jenny Du, and Louis Perez, interview with the author, January 6, 2021.

2. Tom Huddleston Jr., "This Bill Gates–Backed Start-up Is Fighting World Hunger by Making Your Avocados Last Longer," CNBC, December 31, 2018, https://www.cnbc.com/2018/12/31/bill-gates-backed-apeel-sciences -makes-fruit-avocados-last-longer.html.

3. Huddleston, "Bill Gates–Backed Start-up."

4. John Greathouse, "From STEM Researcher to Startup Co-founder—Jenny Du's Venture Is Revolutionizing a $9 Trillion Market," *Forbes*, September 21, 2019, https://www.forbes.com/sites/johngreathouse/2019/09/21/from-stem -researcher-to-startup-co-founder--jenny-dus-venture-is-revolutionizing-a-9 -trillion-market/?sh=1c9b304278ca.

5. Greathouse, "STEM Researcher to Startup Co-founder."

6. Greathouse, "STEM Researcher to Startup Co-founder."

7. Rogers, Du, and Perez, interview, January 6, 2021.

8. Matt Simon, "The Amphiphilic Liquid Coating That Keeps Your Avocados Fresh," Wired, June 21, 2018, https://www.wired.com/story/apeel/.

9. Lauren Barrett and Maya Page, "What's Your Moonshot? Meet the Man Coating Fruits and Vegetables to Alleviate Food Waste and World Hunger," *Newsweek*, August 1, 2019, https://www.newsweek.com/2019/08/09/apeel -sciences-james-rogers-edible-fruit-coating-1452110.html.

10. Barrett and Page, "Moonshot."

11. Kelly Tyko, "Apeel Sciences Has Developed Longer-Lasting Avocadoes, and They're Coming to Stores," *USA Today*, September 18, 2019, https://www .usatoday.com/story/money/food/2019/09/18/apeel-sciences-gives-extra-peel -produce-hopes-reduce-waste/2354066001/.

12. Elaine Watson, "Apeel Sciences Raises $250m to Tackle Food Waste, with Help from Oprah, Katy Perry: 'What We're Selling Is Time,'" FoodNavigator-USA, May 26, 2020, https://www.foodnavigator-usa.com/Article/2020/05 /26/Apeel-Sciences-raises-250m-to-tackle-food-waste-with-invisible-plant -based-skin-with-help-from-Oprah-Katy-Perry?utm_source=newsletter_daily &utm_medium=email&utm_campaign=26-May-2020.

13. Simon, "Amphiphilic Liquid Coating."

14. Tyko, "Apeel Sciences Has Developed Longer-Lasting Avocadoes."

15. Beth Kowitt, "Startup Apeel Is Launching 'Plastic-Free' Cucumbers at Walmart to Cut Back on Waste," *Fortune*, September 21, 2020, https://fortune .com/2020/09/21/apeel-cucumbers-walmart-plastic-food-waste/.

16. Rogers, Du, and Perez, interview, January 6, 2021.

17. Watson, "Apeel Sciences Raises $250m."

18. Apeel Sciences, "What Is Apeel?," accessed March 27, 2021, https://www .apeel.com/faqs.

19. Barrett and Page, "Moonshot."

Chapter 9. Bob Pitzer, Harvest CROO—Picking Strawberries Robotically

1. John Seabrook, "The Age of Robot Farmers," *New Yorker*, April 15, 2019, https://www.newyorker.com/magazine/2019/04/15/the-age-of-robot-farmers.

2. Bob Pitzer, interview with the author, January 10, 2021.

3. Seabrook, "Age of Robot Farmers."

4. Gary Pullano, "Harvest CROO Robotics Strawberry Harvester Nears Fruition," *Fruit Growers News*, March 26, 2019, https://fruitgrowersnews.com /article/harvest-croo-robotics-strawberry-harvester-nears-fruition/.

5. Pullano, "Harvest CROO Robotics Strawberry Harvester."

6. Bob Pitzer, interview with the author, January 10, 2021.

7. Pullano, "Harvest CROO Robotics Strawberry Harvester."

8. University of Florida, Herbert Wertheim College of Engineering, "How an Engineer Picks Strawberries," May 20, 2019, https://www.eng.ufl.edu/newen gineer/alumni-spotlight/pitzer/.

9. Sally French, "How Drones Will Drastically Transform U.S. Agriculture, in One Chart," MarketWatch, November 17, 2015, https://www.marketwatch

.com/story/how-drones-will-drastically-transform-us-agriculture-in-one-chart
-2015-11-17.

10. Danielle Paquette, "Farmworker vs Robot," *Washington Post*, February 17,
2019, https://www.washingtonpost.com/news/national/wp/2019/02/17/
feature/inside-the-race-to-replace-farmworkers-with-robots/.

11. Seabrook, "Age of Robot Farmers."

12. Pullano, "Harvest CROO Robotics Strawberry Harvester."

Chapter 10. Raja Ramachandran, ripe.io—Tracking Food with Blockchain

1. Bernard Marr, "23 Fascinating Bitcoin and Blockchain Quotes Everyone
Should Read," *Forbes*, August 15, 2018, https://www.forbes.com/sites/ber
nardmarr/2018/08/15/23-fascinating-bitcoin-and-blockchain-quotes-every
one-should-read/?sh=4bf087617e8a.

2. Raja Ramachandran, interview with the author, December 8, 2020.

3. Ramachandran, interview, December 8, 2020.

4. Ripe.io, "R3 and ripe.io Partner to Provide Intelligent Transparency and
Trust for the Food and Agricultural Supply Chain on Microsoft Azure," press
release, June 19, 2019, https://www.ripe.io/media/2019/6/18/r3-and-ripeio
-partner-to-provide-intelligent-transparency-and-trust-for-the-food-and
-agricultural-supply-chain-on-microsoft-azure.

5. Ramachandran, interview, December 8, 2020.

6. Mikael Djanian and Nelson Ferreira, "Agriculture Sector: Preparing for
Disruption in the Food Value Chain," McKinsey & Company, April 2020,
https://www.mckinsey.com/~/media/McKinsey/Industries/Agriculture/Our
%20Insights/Agriculture%20sector%20Preparing%20for%20disruption%20
in%20the%20food%20value%20chain/Agriculture-sector-Preparing-for
-disruption-in-the-food-value-chain-vF2.pdf.

7. Ramachandran, interview, December 8, 2020.

8. Ramachandran, interview, December 8, 2020.

9. "Blockchain and IoT for Indoor Farms: Q&A with ripe.io's Raja Ramachandran," Indoor Ag-Con, May 2019, https://indoor.ag/blockchain-and-iot-for
-indoor-farms-q-a-with-ripe-ios-raja-ramachandran/.

10. National Pork Board and Ripe Technology, press release, March 18, 2019.

11. Ramachandran, interview, December 8, 2020.

12. "2019 AgTech Trends: Ripe.io's Inside Scoop on Blockchain," AgTech Salinas,
March 21, 2019, http://agtechsalinasca.com/2019/03/21/2019-agtech-trends
-ripe-ios-inside-scoop-on-blockchain/.

Chapter 11. Lynette Kucsma and Emilio Sepulveda, Foodini—Printing 3D Meals

1. "3D Printing Industry—Worldwide Market Size 2020–2024," Statista Research Department, January 6, 2021, https://www.statista.com/statistics /315386/global-market-for-3d-printers/.

2. Jonathan Chadwick, "Here's How 3D Food Printers Are Changing What We Eat," TechRepublic, November 7, 2017, https://www.techrepublic.com /article/heres-how-3d-food-printers-are-changing-the-way-we-cook.

3. European Institute of Innovation and Technology (EIT), "A 3D Food Printer for Sustainable and Fresh Food," 2019, https://eit.europa.eu/sites/default /files/17.lynette_kucsma.pdf.

4. Jayson Lusk, *Unnaturally Delicious: How Science and Technology Are Serving Up Super Foods to Save the World* (New York: St. Martin's Press, 2016), p. 40.

5. Chadwick, "3D Food Printers Changing What We Eat."

6. Chadwick, "3D Food Printers Changing What We Eat."

7. Donovan Alexander, "3D Printing Will Change the Way You Eat in 2020 and Beyond," Interesting Engineering, March 27, 2020, https://interesting engineering.com/3d-printing-will-change-the-way-you-eat-in-2020-and-beyond.

8. Chadwick, "3D Food Printers Changing What We Eat."

9. Carlota V., "A Guide to 3D Printed Food—Revolution in the Kitchen?," 3DNatives, February 4, 2019, https://www.3dnatives.com/en/3d-printing -food-a-new-revolution-in-cooking/.

10. Emilio Sepulveda, interview with the author, November 19, 2020.

11. Sepulveda, interview, November 19, 2020.

12. Natural Machines, "The Story of Foodini," https://www.naturalmachines.com /about-us.

13. Matt McCue, "Will 3D Printed Food Become as Common as the Microwave?" *Fortune*, February 26, 2015, https://fortune.com/2015/02/26/3d-food -printing/.

14. Jie Sun et al., "An Overview of 3D Printing Technologies for Food Fabrication," *Food and Bioprocess Technology* 8 (2015): 1605–1615, https://doi.org /10.1007/s11947-015-1528-6.

Chapter 12. Daphna Nissenbaum, TIPA—Cutting Plastic Packaging

1. Triodos Investment Management, "'It's a Product and a Movement': Daphna Nissenbaum's Scalable Vision to Tip the Market for Plastics," n.d., accessed

March 27, 2021, https://www.triodos-im.com/articles/2020/tipa-its-a-prod
uct-and-a-movement.

2. Triodos Investment Management, "'It's a Product and a Movement.'"

3. Alessio D'Antino, "Meet Daphna from TIPA, One of the FoodTech 500 Win-
ners," blog post, Forward Fooding, January 21, 2020, https://forwardfooding
.com/blog/blog-category-foodtech500/daphna-from-tipa-foodtech-500.

4. Research and Markets, "Flexible Plastic Packaging Market by Type (Pouches,
Rollstocks, Bags, Wraps), Material (Plastic & Aluminum Foil), Application
(Food, Beverage, Pharma & Healthcare, Personal Care & Cosmetics), Technology, and Region—Global Forecast to 2025," October 2020, https://www
.researchandmarkets.com/reports/5181343/flexible-plastic-packaging-market
-by-type.

5. Daphna Nissenbaum, interview with the author, December 4, 2020.

6. World Economic Forum, "More Plastic than Fish in the Ocean by 2050:
Report Offers Blueprint for Change," news release, January 19, 2016, https://
www.weforum.org/press/2016/01/more-plastic-than-fish-in-the-ocean-by
-2050-report-offers-blueprint-for-change/.

7. World Economic Forum, "More Plastic than Fish in the Ocean."

8. Kovac Family, "Online Shopping with a Clean Conscience," blog post, n.d.,
accessed April 11, 2021, https://kovacfamily.com/blogs/news/online-shop
ping-with-a-clean-conscience.

9. Triodos Investment Management, "'It's a Product and a Movement.'"

10. Kovac Family, "Online Shopping with a Clean Conscience."

11. Nissenbaum, interview, December 4, 2020.

12. Triodos Investment Management, "'It's a Product and a Movement.'"

13. Nissenbaum, interview, December 4, 2020.

14. Triodos Investment Management, "'It's a Product and a Movement.'"

15. Nissenbaum, interview, December 4, 2020.

16. Eleanor Goldberg Fox, "Food Packaging Is Full of Toxic Chemicals—Here's
How It Could Affect Your Health," *Guardian*, May 28, 2019, https://www
.theguardian.com/us-news/2019/may/28/plastics-toxic-america-chemicals
-packaging.

17. D'Antino, "Meet Daphna."

18. Kovac Family, "Online Shopping with a Clean Conscience."

19. Kovac Family, "Online Shopping with a Clean Conscience."

20. Kovac Family, "Online Shopping with a Clean Conscience."

Part 3. Curtail Poisons

1. Ryan Rifai, "UN: 200,000 Die Each Year from Pesticide Poisoning," Aljazeera, March 8, 2017, https://www.aljazeera.com/news/2017/03/200000-die-year-pesticide-poisoning-170308140641105.html.

2. K. L. Bassil et al., "Cancer Health Effects of Pesticides: Systematic Review," *Canadian Family Physician* 53, no. 10 (October 2007): 1704–1711, PMCID: PMC2231435, PMID: 17934034.

3. Joanna Jurewicz and Wojciech Hanke, "Prenatal and Childhood Exposure to Pesticides and Neurobehavioral Development: Review of Epidemiological Studies," *International Journal of Occupational Medicine and Environmental Health* 21, no. 2 (2008): 121–132, https://doi.org/10.2478/v10001-008-0014-z.

4. "How Dangerous Is Pesticide Drift?" *Scientific American*, September 17, 2012, https://www.scientificamerican.com/article/pesticide-drift/.

5. Jeff Tollefson, "Humans Are Driving One Million Species to Extinction," *Nature*, May 6, 2019, https://www.nature.com/articles/d41586-019-01448-4.

6. Elizabeth G. Nielsen and Linda K. Lee, "The Magnitude and Costs of Groundwater Contamination from Agricultural Chemicals: A National Perspective," Resources and Technology Division, Economic Research Service, US Department of Agriculture, Agricultural Economic Report No. 576, 1987, https://naldc.nal.usda.gov/download/CAT88907300/PDF.

7. Sustainable Food Trust, "The True Cost of American Food, April 14–17, 2016, San Francisco: Conference Proceedings," http://sustainablefoodtrust.org/wp-content/uploads/2013/04/TCAF-report.pdf.

8. Rifai, "UN: 200,000 Die Each Year."

9. Kuniyoki Saitoh, Toshira Kuroda, and Seiichi Kumano, "Effects of Organic Fertilization and Pesticide Application on Growth and Yield of Field-Grown Rice for 10 Years," *Japanese Journal of Crop Science* 70, no. 4 (2001): 530–540, https://doi.org/10.1626/jcs.70.530, https://okayama.pure.elsevier.com/en/publications/effects-of-organic-fertilization-and-pesticide-application-on-gro.

10. "Pesticide Industry Propaganda: The Real Story: Myth #8," Environmental Working Group, May 1, 1995, https://www.ewg.org/research/industrys-myths/myth-8-pesticides-cost-money-so-farmers-currently-use-few-pesticides.

Chapter 13. Sébastien Boyer and Thomas Palomares, FarmWise—Plucking Weeds Robotically

1. Adele Peters, "These Giant Robots Are Death Machines for Weeds," Fast Company, May 3, 2019, https://www.fastcompany.com/90342352/these-giant-robots-are-death-machines-for-weeds.

2. Nicole Morell, "Farmers Are Taking Notice of This Robot That Monitors Crops, Pulls Weeds," Slice of MIT, March 28, 2019, https://alum.mit.edu /slice/farmers-are-taking-notice-robot-monitors-crops-pulls-weeds.

3. Morell, "Farmers Are Taking Notice."

4. "FarmWise Unveils Autonomous Vegetable Weeder," *Vegetable Growers News*, December 19, 2018, https://vegetablegrowersnews.com/article/farmwise -unveils-autonomous-vegetable-weeder/.

5. Rachael Lallensack, "Five Roles Robots Will Play in the Future of Farming," *Smithsonian*, September 30, 2019, https://www.smithsonianmag.com/innova tion/five-roles-robots-will-play-future-farming-180973242/.

6. Alexandra Wilson, "FarmWise Raises $14.5 Million to Replace Herbicides with Roving Robots," *Forbes*, September 17, 2019, https://www.forbes.com /sites/alexandrawilson1/2019/09/17/farmwise-raises-14-5-million-to-replace -herbicides-with-roving-robots/#4847155c633b.

7. Michele Catinari, "FarmWise Launches Autonomous Weeding Robot," North America Farm Equipment Magazine, December 4, 2019, https:// www.americafem.com/2019/12/04/farmwise-launches-autonomous-weeding -robot/.

8. Jim Vinoski, "The Farm Automation Breakthrough Bringing the High-Tech West Coast and Rural Rust Belt Together," *Forbes*, May 6, 2019, https://www .forbes.com/sites/jimvinoski/2019/05/06/the-farm-automation-breakthrough -bringing-the-high-tech-west-coast-and-rural-rust-belt-together/#28b86c 2510fe.

9. Vinoski, "Farm Automation Breakthrough."

10. Transparency Market Research, "Agriculture Robots Market," n.d., accessed March 27, 2021, https://www.transparencymarketresearch.com/agriculture -robot-market.html.

11. Matthew Yglesias, "There's Nothing New about Labor-Saving Technology," Slate, February 5, 2013, https://slate.com/business/2013/02/robot-econom ics-are-overblown-labor-saving-technology-isn-t-new.html.

12. B. R. Cohen, "Why Do Visions of Farming's Future Never Involve Farmers?" Slate, June 20, 2019, https://slate.com/technology/2019/06/robot-farming -futurism-precision-agriculture.html.

13. Robot Report Staff, "FarmWise Raises $14.5M Series A for Sustainable Robotic Farming," *Robot Report*, September 17, 2019, https://www.therobot report.com/farmwise-raises-series-a-robotic-weeding/.

14. Wilson, "FarmWise Raises $14.5 Million to Replace Herbicides."

Chapter 14. Jorge Heraud and Lee Redden, Blue River Technology— Spraying Precisely

1. Sonya Mann, "Lettuce-Weeding Robots, Coming Soon to a Farm Near You," *Inc.*, June 16, 2017, https://www.inc.com/sonya-mann/blue-river-technology -ai.html.

2. Mann, "Lettuce-Weeding Robots."

3. Louisa Burwood-Taylor, "How an Artificial Intelligence & Robotics Startup Got Acquired by John Deere," AgFunder Network, October 3, 2017, https:// agfundernews.com/artificial-intelligence-robotics-startup-got-acquired-john -deere.html.

4. Mann, "Lettuce-Weeding Robots."

5. Ben Chostner, "See & Spray: The Next Generation of Weed Control," *Resource*, July/August 2017.

6. Chostner, "See & Spray."

7. Jorge Heraud and Ben Chostner, "Using Robots to Bring Back the 'Magic' of Farms," World Economic Forum, August 27, 2015, https://www.weforum.org /agenda/2015/08/using-robots-to-bring-back-the-magic-of-farms.

8. Burwood-Taylor, "Artificial Intelligence & Robotics Startup."

Chapter 15. Irina Borodina, BioPhero—Messing with Pest Sex

1. Irina Borodina et al., "Fermented Insect Pheromones for Environmentally Friendly Pest Control," paper presented at the International Conference on Industrial Biotechnology and Bioprocessing, August 16–17, 2018, Copenhagen, Denmark, https://www.alliedacademies.org/proceedings/fermented -insect-pheromones-for-environmentally-friendly-pest-control-2572.html.

2. Tiago Pires, "We Can Revolutionise Pest Management," Technologist, January 16, 2019, https://www.technologist.eu/we-can-revolutionise-pest-manage ment/.

3. Pires, "We Can Revolutionise Pest Management."

4. BioPhero, "About Us," https://biophero.com/about/.

5. BioPhero, "About Us."

6. Fortune Business Insights, "Agricultural Pheromones Market Size, Share & COVID-19 Impact Analysis, by Type (Sex Pheromones, Aggregation Pheromones, and Others), Function (Mating Disruption, Detection & Monitoring, and Mass Trapping), Application (Dispensers, Traps, and Spray Method), Crop Type (Field Crops, Orchard Crops, Vegetables, and Others) and Regional Forecast, 2020–2027," January 2021, https://www.fortunebusi nessinsights.com/industry-reports/agricultural-pheromones-market-100071.

7. Grand View Research, "Integrated Pest Management Pheromones Market Size, Share & Trends Analysis Report by Product (Sex, Aggregation, Alarm), by Mode of Application, by Application, by Region, and Segment Forecasts, 2020–2027," April 2020, https://www.grandviewresearch.com/industry -analysis/ipm-pheromones.

8. BioPhero, "About Us."

9. BioPhero, "Insect Resistance," https://biophero.com/the-challenge/insect -resistance/.

10. BioPhero, "Pheromone Solutions," https://biophero.com/pheromone -solutions/.

11. BioPhero, "Customer Demand," https://biophero.com/the-challenge/custo mer-demand/.

Part 4. Nourish Plants

1. Mark Hyman, *Food Fix: How to Save Our Health, Our Economy, Our Communities, and Our Planet—One Bite at a Time* (New York: Little, Brown Spark, 2020).

2. Soil Health Institute, "*Living Soil*: A Documentary for All of Us," https:// livingsoilfilm.com/.

3. R. B. Ferguson, R. M. Lark, and G. P. Slater, "Approaches to Management Zone Definition for Use of Nitrification Inhibitors," *Soil Science Society of America Journal* 67, no. 3 (May 2003): 937–947, https://doi.org/10.2136 /sssaj2003.9370.

Chapter 16. Diane Wu and Poornima Parameswaran, Trace Genomics— Mapping Soils

1. Poornima Parameswaran, interview with the author, March 19, 2020.

2. Parameswaran, interview, March 19, 2020.

3. Parameswaran, interview, March 19, 2020.

4. Trace Genomics, "From Samples to Insights," https://tracegenomics.com /what-we-do.

5. Trace Genomics, "From Samples to Insights."

6. Trace Genomics, "Iowa Farmer, Jerry Dove, on Measuring What's in the Soil," blog post, February 4, 2020, https://tracegenomics.com/iowa-farmer -jerry-dove-on-measuring-whats-in-the-soil/.

7. Foodtank, "From Farms to Incubators: Poornima Parameswaran of Trace Genomics Inc.," September 2019, https://foodtank.com/news/2019/09/from -farms-to-incubators-poornima-parameswaran-of-trace-genomics-inc/.

8. Tom Philpott, *Perilous Bounty: The Looming Collapse of American Farming and How We Can Prevent It* (New York: Bloomsbury, 2020), p. 121.

9. Foodtank, "From Farms to Incubators."

10. Parameswaran, interview, March 19, 2020.

11. Foodtank, "From Farms to Incubators."

Chapter 17. Eric Taipale, Sentera—Analyzing Fields from Above

1. Eric Taipale, interview with the author, May 22, 2020.

2. Louisa Burwood-Taylor, "The Next Generation of Drone Technologies for Agriculture," AgFunder Network, March 16, 2017, https://agfundernews .com/the-next-generation-of-drone-technologies-for-agriculture.html.

3. Neal St. Anthony, "MN Drone Company Sentera Builds Management Team, Raises Funds," *Minnesota Star Tribune*, June 26, 2019, https://www.startribune .com/mn-drone-company-sentera-builds-management-team-raises-funds/511 837322/.

4. McKinsey & Company, "Digital America: A Tale of the Haves and Have-Mores," December 2015, https://www.mckinsey.com/~/media/McKinsey /Industries/Technology%20Media%20and%20Telecommunications/High %20Tech/Our%20Insights/Digital%20America%20A%20tale%20of%20 the%20haves%20and%20have%20mores/MGI%20Digital%20America _Executive%20Summary_December%202015.ashx.

5. Taipale, interview, May 22, 2020.

6. Eric Taipale, interview with the author, December 15, 2020.

7. Burwood-Taylor, "Next Generation of Drone Technologies."

8. Food and Agriculture Organization of the United Nations, "E-Agriculture in Action: Drones for Agriculture," May 4, 2018, http://www.fao.org/in -action/e-agriculture-strategy-guide/documents/detail/en/c/1114182/.

9. Sally French, "How Drones Will Drastically Transform U.S. Agriculture, in One Chart," MarketWatch, November 17, 2015, https://www.marketwatch .com/story/how-drones-will-drastically-transform-us-agriculture-in-one-chart -2015-11-17.

10. Burwood-Taylor, "Next Generation of Drone Technologies."

11. Taipale, interview, December 15, 2020.

12. Taipale, interview, May 22, 2020.

13. Morgan Rose, "Anheuser Busch InBev, Sentera Announce Partnership," *Prairie Star* (Bismarck, ND), January 28, 2020, https://www.agupdate.com /farmandranchguide/news/crop/anheuser-busch-inbev-sentera-announce -partnership/article_1586e4ca-411e-11ea-96ab-53949dfb7d8a.html.

14. Rose, "Anheuser Busch."

15. Taipale, interview, December 15, 2020.

16. Rose, "Anheuser Busch."

17. Taipale, interview, December 15, 2020.

18. Taipale, interview, May 22, 2020.

19. Taipale, interview, May 22, 2020.

20. Taipale, interview, December 15, 2020.

Chapter 18. Ron Hovsepian, Indigo Ag—Providing Probiotics to the Soil

1. Indigo Ag, "Indigo Ag Appoints Ron Hovsepian as CEO," press release, September 2, 2020, https://www.indigoag.com/pages/news/indigo-ag -appoints-ron-hovsepian-as-ceo.

2. Indigo Ag, "Indigo Ag Appoints Ron Hovsepian."

3. Indigo Ag, "Indigo Biological Science," n.d., accessed March 28, 2021, https://www.indigoag.com/biological-science.

4. David R. Montgomery and Anne Biklé, *The Hidden Half of Nature: The Microbial Roots of Life and Health* (New York: W. W. Norton, 2016), p. 239.

5. CBS News, "New Seeds May Help Cotton Farmers in Face of Drought, Climate Change," August 26, 2019, https://www.cbsnews.com/news/new -seeds-may-help-cotton-farmers-in-face-of-drought-climate-change/.

6. David Perry, "Re-imagining Modern Agriculture," blog post, Indigo Ag, September 26, 2017, https://www.indigoag.com/blog/re-imagining-modern -agriculture.

7. Charlie Mitchell, "How an Ag Company Most People Have Never Heard of Could Prove Itself More Disruptive than Netflix or Airbnb," Counter, May 30, 2019, https://thecounter.org/indigo-agriculture-disruptive-startup/.

8. Mitchell, "Ag Company."

9. Mitchell, "Ag Company."

10. Barry Parkin, "Is the Commodity Era Over? Triple Pundit, May 18, 2018, https://www.triplepundit.com/story/2018/commodity-era-over/12266.

11. Lora Kolodny, "The Biggest Breakthrough in Agriculture to Help Feed the Planet May Come from Outer Space," CNBC, May 15, 2019, https://www .cnbc.com/2019/05/15/indigo-ag-improving-yields-with-microbes-satellite -imaging.html.

12. For context, the United States has about five hundred million acres of cultivated land.

13. Mary Cashiola, "Indigo Ag Announces Ambitious Plan to Combat Climate

Change," *Memphis Business Journal,* June 12, 2019, https://www.bizjournals
.com/memphis/news/2019/06/12/indigoag-announces-ambitious-plan-to
-combat.html.

14. Mitchell, "Ag Company."

Chapter 19. Karsten Temme and Alvin Tamsir, Pivot Bio—Feeding Nitrogen to Crop Roots

1. Jennifer Kite-Powell, "Inside the High Tech World of Microbes for Crops," *Forbes,* May 27, 2019, https://www.forbes.com/sites/jenniferhicks/2019/05/27/inside-the-high-tech-world-of-microbes-for-crops/?sh=150c240019ff.

2. Lauren Manning, "Pivot Bio's PROVEN Microbes Can Reduce Negative Impacts of Fertilizer Overuse and Leaching," AgFunder Network, February 7, 2019, https://agfundernews.com/pivot-bios-proven-microbes-can-reduce-negative-impacts-of-fertilizer-overuse.html.

3. Lisa M. Krieger, "Berkeley's Designer Bacteria: The End of Fertilizer?," *San Jose (CA) Mercury News,* March 5, 2019, https://www.mercurynews.com/2019/03/05/berkeleys-designer-bacteria-the-end-of-fertilizer/.

4. Karsten Temme, "The Crop Microbiome Holds the Future of Fertilizer," blog post, Pivot Bio, February 13, 2018, https://blog.pivotbio.com/the-crop-microbiome-holds-the-future-of-fertilizer.

5. Climatewire, "How Fertilizer Is Making Climate Change Worse," September 16, 2020, https://www.eenews.net/climatewire/stories/1063713787.

6. Andrew Zaleski, "The Corn That Grows Itself," OneZero, June 13, 2019, https://onezero.medium.com/how-microbes-could-upend-americas-toxic-dependence-on-nitrogen-fertilizer-548451117a63.

7. Kite-Powell, "High Tech World of Microbes."

8. Karsten Temme, interview with the author, January 13, 2021.

9. Kite-Powell, "High Tech World of Microbes."

10. Temme, interview, January 13, 2021.

11. Temme, interview, January 13, 2021.

12. Temme, interview, January 13, 2021.

13. Zaleski, "Corn That Grows Itself."

14. Zaleski, "Corn That Grows Itself."

15. Manning, "Pivot Bio's PROVEN Microbes."

16. Temme, interview, January 13, 2021.

17. Manning, "Pivot Bio's PROVEN Microbes."

18. Zaleski, "Corn That Grows Itself."

19. Krieger, "Berkeley's Designer Bacteria."

Chapter 20. Tony Alvarez, WaterBit—Watering Precisely

1. Mark Trapolino, "New Tech Overcomes Agricultural Challenges," Fierce-Electronics, March 5, 2019, https://www.fierceelectronics.com/components /new-tech-overcomes-agricultural-challenges.

2. Dana Marotto, "WaterBit Precision Irrigation Solution Adds Major Enhancements," WaterBit press release, February 11, 2020, https://www.waterbit.com /waterbit-precision-irrigation-solution-adds-major-enhancements/.

3. WaterBit, "Experienced IoT and Semiconductor Executive to Lead Agtech Startup," press release, April 2, 2020, https://www.waterbit.app/waterbit-tony -alvarez-ceo/.

4. Tony Alvarez, interview with the author, April 2, 2020.

5. Nathalee Ghafouri, "Tech Note: Growing Premium Wine Grapes in the Living Lab at Clos de la Tech Winery," WaterBit, March 25, 2020, https://www .waterbit.com/premium-wine-grapes-waterbit-clos-de-la-tech/.

6. Jeff Caldwell, "Innovation in Irrigation," *FarmLife*, Spring 2020, https:// myfarmlife.com/asides/innovation-in-irrigation/.

7. Caldwell, "Innovation in Irrigation."

Part 5. Cut Carbon

1. Kim Severson, "From Apples to Popcorn, Climate Change Is Altering the Foods America Grows," *New York Times*, April 30, 2019, https://www .nytimes.com/2019/04/30/dining/farming-climate-change.html.

2. Jonathan Knutson, "U.S. Organic Market Tops $50 Billion," Agweek, June 10, 2019, https://www.agweek.com/business/4622665-us-organic-market -tops-50-billion.

3. Rachel Cernansky, "We Don't Have Enough Organic Farms. Why Not?" *National Geographic*, November 20, 2018, https://www.nationalgeographic .com/environment/future-of-food/organic-farming-crops-consumers/#.

4. "Got Questions? Ask the Green Geek," *Pima County (AZ) FYI Newsletter* 4, no. 24 (June 14, 2019), https://webcms.pima.gov/cms/One.aspx?portalId =169&pageId=490799.

5. Matthew N. Hayek and Rachael D. Garrett, "Nationwide Shift to Grass-Fed Beef Requires Larger Cattle Population," *Environmental Research Letters* 13, no. 8 (July 25, 2018): 084005, https://doi.org/10.1088/1748-9326/aad401.

6. Janet Ranganathan et al., "Regenerative Agriculture: Good for Soil Health,

but Limited Potential to Mitigate Climate Change," blog post, World Resources Institute, May 12, 2020, https://www.wri.org/blog/2020/05/regen erative-agriculture-climate-change.

Chapter 21. Rachel Haurwitz, Caribou Biosciences—Editing Genes

1. Stephen S. Hall, "Crispr Can Speed Up Nature—and Change How We Grow Food," Wired, July 17, 2018, https://www.wired.com/story/crispr-tomato -mutant-future-of-food/.

2. Todd Funke et al., "Molecular Basis for the Herbicide Resistance of Roundup Ready Crops," *Proceedings of the National Academy of Sciences of the United States of America* 103, no. 35 (2006): 13010–13015, https://doi.org/10.1073 /pnas.0603638103.

3. Jorge Fernandez-Cornejo et al., "Genetically Engineered Crops in the United States," US Department of Agriculture Economic Research Report No. 162, February 2014, https://www.ers.usda.gov/webdocs/publications/45179 /43668_err162.pdf?v=2759.9.

4. David Baltimore et al., "CRISPR Controversy," Translational Scientist, January 21, 2016, https://thetranslationalscientist.com/tools-techniques/crispr -controversy.

5. Andrew Pollack, "Jennifer Doudna, a Pioneer Who Helped Simplify Genome Editing," *New York Times*, May 11, 2015, https://www.nytimes.com/2015/05 /12/science/jennifer-doudna-crispr-cas9-genetic-engineering.html.

6. Molly Fosco, "This Scientist Turned CEO Wants to Gene-Edit a Way to Cure Cancer," Ozy, March 15, 2018, https://www.ozy.com/the-new-and-the-next /this-scientist-turned-ceo-wants-to-gene-edit-a-way-to-cure-cancer/85243/.

7. Fosco, "Scientist Turned CEO."

8. Fosco, "Scientist Turned CEO."

9. Rachel Haurwitz, interview with the author, December 10, 2020.

10. Eric Niiler, "Why Gene Editing Is the Next Food Revolution," *National Geographic*, August 10, 2018, https://www.nationalgeographic.com/environment /article/food-technology-gene-editing.

11. Dyllan Furness, "From Corn to Cattle, Gene Editing Is About to Supercharge Agriculture," Digital Trends, April 17, 2017, https://www.digitaltrends.com /cool-tech/crispr-gene-editing-and-the-dna-of-future-food/.

12. Niiler, "Gene Editing."

13. US Department of Agriculture, "Secretary Perdue Issues USDA Statement on Plant Breeding Innovation," March 28, 2018, https://www.usda.gov/media

/press-releases/2018/03/28/secretary-perdue-issues-usda-statement-plant -breeding-innovation.

14. Niiler, "Gene Editing."

15. Niiler, "Gene Editing."

16. Hall, "Crispr Can Speed Up Nature."

17. Hall, "Crispr Can Speed Up Nature."

18. Washington Post Live, "Transcript: A Conversation with Jennifer Doudna and Walter Isaacson," *Washington Post*, March 15, 2021, https://www.wash ingtonpost.com/washington-post-live/2021/03/15/transcript-conversation -with-jennifer-doudna-walter-isaacson/.

19. Niiler, "Gene Editing."

Chapter 22. Lee DeHaan, The Land Institute—Planting Perennials

1. Kathryn Shattuck, "Q. and A.: Farming for an Uncertain Future," *New York Times*, September 27, 2012, https://green.blogs.nytimes.com/2012/09 /27/q-and-a-farming-for-an-uncertain-future/.

2. Land Institute, "Lee DeHaan," https://landinstitute.org/about-us/staff/lee -dehaan/.

3. David Pimentel et al., "Economic and Environmental Benefits of Biodiversity," *BioScience* 47, no. 11 (December 1997): 747–757, https://doi.org/10 .2307/1313097.

4. Lee DeHaan, interview with the author, November 12, 2020.

5. DeHaan, interview, November 12, 2020.

6. Jerry D. Glover and John P. Reganold, "Perennial Grains: Food Security for the Future," *Issues in Science and Technology* 26, no. 2 (Winter 2010), https:// issues.org/glover/.

7. DeHaan, interview, November 12, 2020.

8. Don Wyse, interview with the author, September 10, 2019.

9. Jerry Steiner, interview with the author, September 10, 2019.

10. Daniel Cusick, "Grain May Take a Big Bite out of Cropland Emissions," E&E News, May 7, 2019, https://www.eenews.net/stories/1060290955/print.

11. Cusick, "Grain May Take a Big Bite."

12. DeHaan, interview, November 12, 2020.

13. Land Institute, "Perennial Crops: New Hardware for Agriculture," n.d., accessed March 28, 2021, https://landinstitute.org/our-work/perennial-crops/.

14. DeHaan, interview, November 12, 2020.

Chapter 23. Joshua Goldman, Australis Aquaculture—Blocking Burps

1. Estimates vary. *World Watch Magazine* calculates that livestock accounts for 51 percent of worldwide greenhouse-gas emissions. Robert Goodland and Jeff Anhang, "Livestock and Climate Change," *World Watch Magazine*, November/December 2009, https:awellfedworld.org/wp-content/uploads/Livestock-Climate-Change-Anhang-Goodland.pdf. According to the Food and Agriculture Organization of the United Nations, 37 percent of human-induced methane comes from livestock. The US Environmental Protection Agency says enteric fermentation by beef cattle accounts for 2 percent of the nation's greenhouse-gas emissions.

2. Quote by Elizabeth Latham, cofounder of Bezoar Laboratories, in Thin Lei Win, "Fighting Global Warming, One Cow Belch at a Time," Reuters, July 19, 2018, https://www.reuters.com/article/us-global-livestock-emissions-idUSKBN1K91CU.

3. Barry Estabrook, "The Anti-salmon: A Fish We Can Finally Farm without Guilt," *Atlantic*, October 12, 2010, https://www.theatlantic.com/health/archive/2010/10/the-anti-salmon-a-fish-we-can-finally-farm-without-guilt/64317/.

4. Australis, "Australis Launches Greener Grazing: 'Aquatic Moonshot' to Combat Climate Change," blog post, n.d., accessed April 3, 2021, https://www.thebetterfish.com/thecurrent/greener-grazing/.

5. Australis, "Australis Launches Greener Grazing."

6. Breanna Michell Roque et al., "Effect of the Macroalgae *Asparagopsis taxiformis* on Methane Production and Rumen Microbiome Assemblage," *Animal Microbiome* 1, no. 3 (2019), https://doi.org/10.1186/s42523-019-0004-4.

7. Joshua Goldman, interview with the author, November 16, 2020.

8. Xixi Li et al., "*Asparagopsis taxiformis* Decreases Enteric Methane Production from Sheep," *Animal Production Science* 58, no. 4 (September 28, 2016): 681–688, https://doi.org/10.1071/AN15883. See also B. M. Roque et al., "Red Seaweed (*Asparagopsis taxiformis*) Supplementation Reduces Enteric Methane by over 80 Percent in Beef Steers," *bioRxiv*, July 16, 2020, https://doi.org/10.1101/2020.07.15.204958.

9. Goldman, interview, November 16, 2020.

10. Algae World News, "The Fish Farmer Growing Seaweed to Feed Cows and Save the Planet," February 17, 2020, https://news.algaeworld.org/2020/02/the-fish-farmer-growing-seaweed-to-feed-cows-and-save-the-planet/.

11. Goldman, interview, November 16, 2020.

12. Goldman, interview, November 16, 2020.

13. Goldman, interview, November 16, 2020.

14. D. Van Wesemael et al., "Reducing Enteric Methane Emissions from Dairy Cattle: Two Ways to Supplement 3-Nitrooxypropanol," *Journal of Dairy Science* 102, no. 2 (February 2019): 1780–1787, https://doi.org/10.3168 /jds.2018-14534.

15. Breanna M. Roque et al., "Effect of Mootral—a Garlic- and Citrus-Extract-Based Feed Additive—on Enteric Methane Emissions in Feedlot Cattle," *Translational Animal Science* 3, no. 4 (August 16, 2019): 1383–1388, https:// doi.org/10.1093/tas/txz133.

16. As much as 12 percent of a cow's energy intake is used to produce and emit methane. Adam Satariano, "The Business of Burps: Scientists Smell Profit in Cow Emissions," *New York Times*, May 1, 2020, https://www.nytimes.com /2020/05/01/business/cow-methane-climate-change.html.

17. "Mootral: Reducing GHG Emissions through Innovative Solutions," Swiss Food & Nutrition Valley, February 2021, https://swissfoodnutritionvalley.ch /mootral-reducing-ghg-emissions-through-innovative-solutions/.

18. Rowan Walrath, "The Answer to Climate-Killing Cow Farts May Come from the Sea," *Mother Jones*, October 27, 2019, https://www.motherjones.com /food/2019/10/methane-climate-change-cows-seaweed-australis-kebreab-asparagopsis/.

Chapter 24. Julia Collins, Planet FWD—Creating a Climate-Friendly Food Platform

1. Mark Noack, "Zume Pizza Lays Off 172 Workers in Mountain View," *Mountain View (CA) Voice*, January 17, 2020, p. 5.

2. John Ketchum, "Julia Collins Turned Zume Pizza into a $2 Billion Company. Now, She's Trying to Save the World," AfroTech, September 12, 2019, https://afrotech.com/julia-collins-transformed-the-food-industry-now-shes -trying-to-save-the-world.

3. Julia Collins, interview with the author, March 2, 2021.

4. Megan Rose Dickey, "Zume Co-founder Goes from Pizza to Climate-Friendly Food with $2.7 Million in Funding," TechCrunch, March 12, 2020, https:// techcrunch.com/2020/03/12/zume-co-founder-goes-from-pizza-to-climate -friendly-food-with-2-7-million-in-funding.

5. Jules Pretty et al., "Global Assessment of Agricultural System Redesign for Sustainable Intensification," *Nature Sustainability* 1 (August 14, 2018): 441– 446, https://doi.org/10.1038/s41893-018-0114-0.

6. Ketchum, "Julia Collins."

7. Ketchum, "Julia Collins."

8. Gené Teare, "2018 Sets All-Time High for Investment Dollars into Female-Founded Startups," Crunchbase News, January 15, 2019, https://news.crunchbase.com/news/2018-sets-all-time-high-for-investment-dollars-into-female-founded-startups/#:~:text=Global%20Venture%20Deal%20Volume,venture%20rounds%20with%20known%20founders.

9. Ann-Derrick Gaillot, "The Venture Capital World Has a Problem with Women of Color," girlboss, January 21, 2019, https://www.girlboss.com/read/venture-capital-woc-women-of-color.

10. Ketchum, "Julia Collins."

11. "Portfolio: Planet FWD," Kapor Capital, n.d., accessed March 28, 2021, https://www.kaporcapital.com/portfolio/planet-fwd/.

Chapter 25. Stafford Sheehan and Gregory Constantine, Air Company—Cutting Carbon with Vodka

1. Gregory Constantine, interview with the author, January 19, 2021.

2. Constantine, interview, January 19, 2021.

3. Adele Peters, "This Carbon-Negative Vodka Is Made from Captured CO2," Fast Company, November 7, 2019, https://www.fastcompany.com/90422270/this-carbon-negative-vodka-is-made-from-captured-co2.

4. Peters, "Carbon-Negative Vodka."

5. Afdhel Aziz, "The Power of Purpose: How Air Co. Is Tackling Climate Change by Making Products That Are Out of This World," Forbes, January 9, 2020, https://www.forbes.com/sites/afdhelaziz/2020/01/09/the-power-of-purpose-how-air-co-is-tackling-climate-change-by-making-vodka-that-is-out-of-this-world/#324b101b2525.

6. Air Company website, accessed March 28, 2021, https://aircompany.com.

7. Aziz, "Power of Purpose."

8. Darrell Etherington, "CO2-Based Vodka Startup Air Co. Fully Redirects Its Tech to Making Hand Sanitizer for Donation," TechCrunch, March 17, 2020, https://techcrunch.com/2020/03/17/co2-based-vodka-startup-air-co-fully-redirects-its-tech-to-making-hand-sanitizer-for-donation/.

9. Peters, "Carbon-Negative Vodka."

10. Amy Hopkins, "The Luxury Masters 2019 Result," Spirits Business, October 1, 2019, https://www.thespiritsbusiness.com/2019/10/the-luxury-spirits-masters-2019-results/#:~:text=In%20the%20Vodka%20%E2%80%93%2020Ultra%20Premium%20round%2C%20the,little%20viscosity%2C%E2%80%9D%20said%20Roos%20about%20Air%20Co%20Vodka.

11. John Kell, "The Unconventional Methods Liquor Makers Are Taking to Be More Sustainable," *Fortune*, November 7, 2019, https://fortune.com/2019/11/07/sustainable-spirits/.

Conclusion. Disrupting Farms and Foods

1. Jonathan Safran Foer, *Eating Animals* (New York: Little, Brown, 2009), pp. 103–104.

2. Carol M. Kopp, "Creative Destruction," Investopedia, updated February 12, 2021, https://www.investopedia.com/terms/c/creativedestruction.asp.

3. Amanda Little, interview with the author, September 16, 2019.

4. Peter H. Diamandis and Steven Kotler, *The Future Is Faster than You Think: How Converging Technologies Are Transforming Business, Industries, and Our Lives* (New York: Simon & Schuster, 2020), p. 23.

5. MarketsandMarkets, "Dairy Alternatives Market by Source (Soy, Almond, Coconut, Oats, Rice, Hemp), Application (Milk, Yogurt, Ice Creams, Cheese, Creamers), Distribution Channel (Supermarkets, Health Food Stores, Pharmacies), Formulation, and Region—Global Forecast to 2026," January 2021, https://www.marketsandmarkets.com/Market-Reports/dairy-alternative-plant-milk-beverages-market-677.html.

6. Marian Bull, "Still Got It?" *New York Times*, March 15, 2020, p. ST1.

7. Andrew Jacobs, "Facing an Anti-dairy Movement, Farms Alter Practices," *New York Times*, December 29, 2020, p. D8.

8. Paul Shapiro, *Clean Meat: How Growing Meat without Animals Will Revolutionize Dinner and the World* (New York: Gallery Books, 2018), preface.

9. Harold Evans, with Gail Buckland and David Lefer, *They Made America: From the Steam Engine to the Search Engine; Two Centuries of Innovators* (New York: Little, Brown, 2004), p. 11.

10. Jules Billard, "The Revolution in American Agriculture," *National Geographic*, February 1970.

11. Jonathan Safran Foer, *We Are the Weather: Saving the Planet Begins at Breakfast* (New York: Farrar, Straus and Giroux, 2019), p. 208.

12. Rachel Laudan, "A Plea for Culinary Modernism: Why We Should Love New, Fast, Processed Food," *Gastronomica* 1, no. 1 (February 2001): 36–44, https://doi.org/10.1525/gfc.2001.1.1.36.

13. David Wallace-Wells, *The Uninhabitable Earth: Life after Warming* (New York: Tim Duggan Books, 2019), p. 57.

14. Eric Asimov, "In Oregon Wine Country, One Farmer's Battle to Save the

Soil," *New York Times*, October 17, 2019, https://www.nytimes.com/2019/10/17/dining/drinks/climate-change-regenerative-agriculture-wine.html.

15. In Kind, "The Impossible Burger and Earth's Future," Medium, June 8, 2018, https://medium.com/in-kind/the-impossible-burger-and-earths-future-c6769 99f1e13.

16. Deborah White, "What Are U.S. Farm Subsidies?" ThoughtCo, May 29, 2020, https://www.thoughtco.com/us-farm-subsidies-3325162.

17. Dan Charles, "Farmers Got Billions from Taxpayers in 2019, and Hardly Anyone Objected," NPR, December 31, 2019, https://www.npr.org/sections/thesalt/2019/12/31/790261705/farmers-got-billions-from-taxpayers-in-2019-and-hardly-anyone-objected.

18. Sharon LaFraniere, "More Billions to Farms, This Time for Virus Aid," *New York Times*, June 8, 2020, p. A1.

19. Alan Rappeport, "$46 Billion Heads Farmers' Way as Trump Seeks to Bolster Base," *New York Times*, October 13, 2020, p. A1.

20. Environmental Working Group, "The United States Farm Subsidy Information," n.d., accessed March 29, 2021, https://farm.ewg.org/region.php?fips=00000&statename=UnitedStates.

21. Environmental Working Group, "Commodity Subsidies in the United States Totaled $240.5 Billion from 1995–2020," n.d., accessed March 29, 2021, https://farm.ewg.org/progdetail.php?fips=00000&progcode=totalfarm&page=conc&regionname=theUnitedStates.

22. American Farm Bureau Federation, "Farm Bankruptcies Rise Again," Market Intel, October 30, 2019, https://www.fb.org/market-intel/farm-bankruptcies-rise-again.

23. Eric Kessler, interview with the author, November 11, 2020.

24. Frank Viviano, "This Tiny Country Feeds the World," *National Geographic*, September 2017.

25. Jan Conway, "U.S. Plant-Based Milks—Statistics & Facts," Statista, December 1, 2020, https://www.statista.com/topics/3072/us-plant-based-milks/.

26. Louisa Burwood-Taylor, "How an Artificial Intelligence & Robotics Startup Got Acquired by John Deere," AgFunder Network, October 3, 2017, https://agfundernews.com/artificial-intelligence-robotics-startup-got-acquired-john-deere.html.

27. Ariel Schwartz, "The $325,000 Lab-Grown Hamburger Now Costs Less than $12," Fast Company, April 1, 2015, https://www.fastcompany.com/3044572/the-325000-lab-grown-hamburger-now-costs-less-than-12.

About the Author

Richard Munson is the author of several books, most recently *Tesla: Inventor of the Modern.* He also has written a biography of the undersea explorer and filmmaker Jacques Cousteau, a history of electricity, and a behind-the-scenes look at how congressional appropriators spend taxpayer money. Now based in the Chicago area, he has worked on environmental and clean-energy issues at nonprofits, within universities, in the private sector, and on Capitol Hill.

Index